U0252179

CCICED Policy Research Report on Environment and Development 2018

2018

CHINA COUNCIL FOR INTERNATIONAL COOPERATION ON ENVIRONMENT AND DEVELOPMENT POLICY RESEARCH REPORT ON ENVIRONMENT AND DEVELOPMENT

INNOVATION FOR A GREEN NEW ERA

Edited by
China Council for International Cooperation on Environment and Development Secretariat

China Environment Publishing Group·Beijing

图书在版编目（CIP）数据

中国环境与发展国际合作委员会环境与发展政策研究报告.2018，创新引领绿色新时代 = CHINA COUNCIL FOR INTERNATIONAL COOPERATION ON ENVIRONMENT AND DEVELOPMENT POLICY RESEARCH REPORT ON ENVIRONMENT AND DEVELOPMENT 2018:INNOVATION FOR A GREEN NEW ERA：英文 / 中国环境与发展国际合作委员会秘书处编．-- 北京：中国环境出版集团，2019.6
ISBN 978-7-5111-3974-0

Ⅰ．①中… Ⅱ．①中… Ⅲ．①环境保护－研究报告－中国－2018－英文 Ⅳ．①X-12

中国版本图书馆 CIP 数据核字（2019）第 085328 号

出 版 人　武德凯
责任编辑　黄　颖　张秋辰
责任校对　任　丽
装帧设计　宋　瑞

出版发行　中国环境出版集团
（100062　北京市东城区广渠门内大街 16 号）
网　　址：http://www.cesp.com.cn
电子邮箱：bjgl@cesp.com.cn
联系电话：010-67112765（编辑管理部）
发行热线：010-67125803，010-67113405（传真）
印　　刷　北京建宏印刷有限公司
经　　销　各地新华书店
版　　次　2019 年 6 月第 1 版
印　　次　2019 年 6 月第 1 次印刷
开　　本　787×960　1/16
印　　张　10
字　　数　165 千字
定　　价　60.00 元

Note on this volume

China Council for International Cooperation on Environment and Development (CCICED) held the 2018 Annual General Meeting (AGM) with the theme of "Innovation for a Green New Era" from November 1st to 3rd. Mr. Han Zheng, Vice Premier of the State Council and Chairperson of CCICED delivered a keynote speech "Persist in Reform and Innovation to Inject New Impetus into Green Development" at the opening session of the conference. The conference reviewed the policy researches under the Task Forces (TFs) on "Global Governance and Ecological Civilization", "Green Urbanization and Environmental Improvement", "Innovation, Sustainable Production and Consumption" and "Green Energy, Investment and Trade". It also included the roundtable discussion "Low-carbon and Circular Economy for Green Development" and the eight open forums with themes of "Green Belt and Road and 2030 Agenda for Sustainable Development", "Addressing Climate Change by Innovative Development Pathways", "Post 2020 Biodiversity Conservation", "Global Marine Environmental Governance", "Innovation-Driven Green Urbanization", "Beautiful China 2035", "Green Consumption for Green Transformation" and "Green Development Innovation for the Yangtze River Economic Belt", and gathered advices from Chinese and international members and experts, finally forming a series of Policy Recommendations to be submitted to the Chinese government.

The CCICED Council Members believed that the world would be confronted with various impacts, most of which were related to environmental risks. Additionally, some major powers are exiting from environmental and other international agreements, trade war may go against the actions to deal with global climatologic change while biodiversity loss would be adverse to poverty alleviation. Currently, China has become a leader in promoting global green development, and the new challenge that it faces is advancing its domestic green development, and becoming a leader in global green development. The Chinese government will be dedicated to achieving its "beautiful China" objective, building a healthy and beautiful earth for our descendants.

This report covers the 2018 policy research findings of CCICED, the 2018 Policy Recommendations submitted to the Chinese government, "Progress on Environment and Development Policies in China (2017—2018) and Impact of CCICED's Policy Recommendations", and "CCICED 2018 Issues Paper", for reference by decision-makers at all levels, experts, scholars and the public.

The Chinese and International experts and other individuals who prepared each document are listed below:

Chapter 1

LIU Hui, John MIMIKAKIS, CAO Ling, HAN Yang, John VIRDIN, Jake KRITZER, SUN Fang

Chapter 2

LIU Shijin, Arthur HANSON, ZHANG Yongsheng, LI Yonghong, ZHANG Jianyu, Knut ALFSEN, Dimitri de Boer, WANG Yong, ZHANG Huiyong, QIN Hu

Chapter 3

LIU Shijin, Arthur HANSON, ZHANG Yongsheng, LI Yonghong, ZHANG Jianyu, Knut ALFSEN, Dimitri de Boer, WANG Yong, ZHANG Huiyong, QIN Hu

Chapter 4

LIU Shijin, Arthur HANSON, ZHANG Yongsheng, LI Yonghong, ZHANG Jianyu, Knut ALFSEN, Dimitri de Boer, WANG Yong, ZHANG Huiyong, QIN Hu

Comprehensively Promote the Reform and Innovation to Provide a Strong Guarantee for the Fight against Pollution*

— Keynote speech by Minister Li Ganjie at the 2018 Annual General Meeting (AGM) of the China Council for International Cooperation on Environment and Development (CCICED)

This is the first CCICED AGM hosted by the newly established Ministry of Ecology and Environment(MEE). The Chinese government has always attached great importance to the environment. This year celebrates the 40th anniversary of reform and opening up – a critical phase in China's history.

The Chinese government has passed policies that indicate China's path to advanced development. In May, President Xi Jinping delivered an important speech at the national environmental conference in Beijing. He discussed ecological civilization, equal prosperity, the harmonious coexistence of humans and nature, a stable path to universal well-being, and an insistence on the most stringent protection of the environment.

In June, the Chinese government issued a call for strengthening environmental protection and resolutely fighting pollution. Regarding institutional reform, the new MEE will focus on setting standards for environmental policies.

We have rolled out a blue-sky protection campaign, identifying key areas to treat air pollution issues in the winter season this year. We have targeted campaigns against the illegal transit and dumping of solid and dangerous waste.

By the end of the year, we will complete red-line mapping in 16 provinces, which will be a significant policy measure to protect the environment. The overall blue-sky day counts rose by 1.3%. The density of $PM_{2.5}$ was cut by 9.8% and PM_{10} by 5.6%. In Beijing, the concentration of $PM_{2.5}$ has dropped by 16.7% to 15 micrograms per cubic metre, and PM_{10} by 23%. There have also been improvements to water, with increases in top grade water and decreases in the more polluted grades of water. While we have seen success, we also face further challenges. Air pollution has been more severe.

* This is the transcript of the speech by Mr. Li Ganjie, Minister of Ecology and Environment, and Executive Vice Chair of CCICED, at the opening session of 2018 AGM.

China will promote broad green development concepts, ecological protection red lines, as well as caps on natural resource utilization. We will implement vertical as well as horizontal supervision and enforcement to ensure the effectiveness of environmental enforcement.

China will be proactive in supporting a green-innovation system, to allow for crucial market breakthroughs in the treatment of pollution. The government will continue to implement water body treatment controls and research. Along with setting up an awards system for whistleblowers, China will maintain public campaigns to raise awareness of environmental issues.

Contents

Chapter 1

Special Policy Study on Global Ocean Governance and Ecological Civilization[1]

1 This chapter is contributed by the TT2 on Marine Resources and Biodiversity under the SPS on Ocean Governance co-chaired by Mr. Jan-Gunnar Winther and Mr. Su Jilan.

1.1 Introduction

1.1.1 The value of living marine resources

Ocean ecosystems harbor immense biodiversity, an abundance of life that provides numerous benefits for people. These living marine resources have cultural value in many places, playing a central role in mythology, religion, symbols and stories worldwide, from antiquity right up to modern times. Many people enjoy directly interacting with ocean life through scuba diving, birdwatching, sport fishing and other forms of recreation, and these activities support a lucrative global tourism industry. Living marine resources also provide important regulating and supporting services, such as oxygen production, nutrient cycling, water filtration, carbon sequestration and storm buffering. The value of these services to a growing ocean economy is difficult to estimate, but is likely on the order of hundreds of billions, if not trillions, of U.S.dollars worldwide each year[1]. Yet, the full breadth of services provided by living marine resources is not well appreciated. Consequently too many policies fail to safeguard living marine resources, biodiversity is in decline and ecosystem services and the value they provide are being lost; this poses a risk to human health, food security, poverty and livelihoods.

Perhaps the most significant and widespread ecosystem service delivered by life in the oceans is provision of seafood through wild capture fisheries and mariculture. For some, seafood makes for a more interesting and varied diet and provides a deeper connection to the sea. For others, especially in many parts of the developing tropics, seafood is a critical component of food and nutrition security. Indeed, 20% of the world's population is critically dependent upon seafood as a source of micronutrients, so called because only small amounts are required to fulfill vital physiological functions[2]. Fisheries and mariculture not only provide food, but income and livelihood, to 13.8 million fisheries practitioners in China. These benefits are not limited solely to those who harvest food from the sea, but also extend to the many support services—gear manufacturers, vessel mechanics, etc.—as well as processors, restaurants and other businesses along the supply chain that are also built on seafood. The value of these other seafood-dependent industries can substantially add to the employment and revenue generated by the production sectors.

These promising findings notwithstanding, most future increases in the global supply

1 DE GROOT R, BRANDER L, PLOEG S V, et al. Global estimates of the value of ecosystems and their services in monetary units[J/OL]. Ecosystem Services,2012,1(1), 50-61. doi:10.1016/j.ecoser.2012.07.005.
2 GOLDEN C D, ALLISON E H, CHEUNG W W, et al. Nutrition: Fall in fish catch threatens human health[J/OL]. Nature, 2016,534(7607), 317-320. doi:10.1038/534317a.

of seafood will come from mariculture production. The volume of seafood generated by mariculture has long been on the rise, and likely will soon equal and surpass production from wild capture fisheries. Moreover, there is considerable potential for additional mariculture production worldwide[1].

On the other hand, mariculture can also pose considerable risks to marine ecosystems and other values generated by living marine resources, including wild capture fisheries. For example, large-scale infestations of green algae *Ulva prolifera* off the coast of the northern Yellow Sea have occurred repeatedly in recent years, threatening the valuable tourism industry in the region. While multiple factors contributed to these infestations, the most important driver has been identified as coastal seaweed farms from as far away as Jiangsu Province that provided substantial additional surface area for algal growth [2, 3].

Of course mariculture can also produce ecological benefits, including water filtration and creation of nursery habitat by seaweed aquaculture[4], but these benefits are typically ancillary rather than planned or optimized.

1.1.2 China's living marine resources

The exclusive economic zone (EEZ) of China covers a wide coastal ocean area from a coastline that spans 18,000 kilometers across more than 20° of latitude. This EEZ stretches from the tropical waters of the Beibu Gulf and South China Sea, through the sub-tropical East China Sea, and into the temperate Yellow Sea in the north. China's EEZ and the Bohai Sea represent three of the world's large marine ecosystems (LMEs), and their pronounced biogeographic gradient means that the diversity of living marine resources in China is substantial. Indeed, the ocean economy, including the value generated by living marine resources and other aspects of marine ecosystems, has become an important part of China's economic growth, generating more than 700 billion yuan annually, or more than 9% of GDP in 2017[5].

1 GENTRY R, FROEHLICH H, GRIMM D, et al. Mapping the global potential for marine aquaculture[J]. Nature Ecology & Evolution, 2017,1: 1317-1324.

2 PANG S J, LIU F, SHAN T F, et al. Tracking the algal origin of the Ulva bloom in the Yellow Sea by a combination of molecular, morphological and physiological analyses[J]. Marine environmental research, 2010,69(4):207-215. Wang W, Liu H, Li YQ, et al. Development and management of land reclamation in China. Ocean and Coastal Management,2014, 102: 415-425.

3 LIU D, KEESING J K, HE P, et al. The world's largest macroalgal bloom in the Yellow Sea, China: Formation and implications. Estuarine[J], Coastal and Shelf Science,2013, 129: 2-10.

4 Liu H, Sun L, Wang J, et al. Current status, problems and countermeasures of environmental-friendly mariculture.//Tang Q.Environment-Friendly Mariculture Development Strategy: New Ideas, New Tasks, New Approaches [M] Beijing:Science Press,2017.

5 State Oceanic Administration. 2017 China Marine Economic Statistics Bulletin[J/OL]. 2018.http://www.cme.gov.cn//node/434.jspx.

Of the many values provided by living marine resources, China today depends most heavily on domestic production of seafood, which fulfills most of its demand (Figure 1-1). Chinese seafood production is consumed largely within China, accounting for 85% ~ 95% of domestic aquatic product consumption. The export volume of aquatic products is only around 6% of total production; however, China's seafood export is at the high-value end. In 2016, the export value of aquatic products was $20.7 billion, or 28% of China's gross export of agricultural products. By comparison, the import value of aquatic products was $9.4 billion, much of which was low-value fishmeal[1].

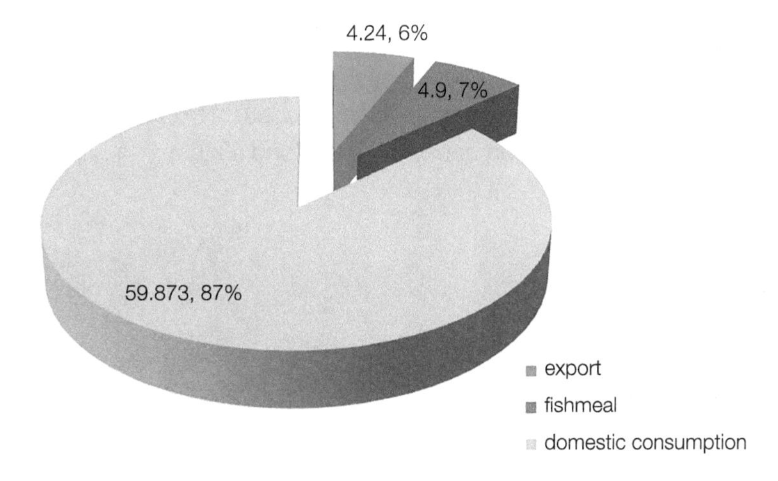

Figure 1-1 Markets for Aquatic Products in China, 2016 (data in million metric tons and percentage)[2]

China's demand for seafood and other high-value marine products is growing and China is increasingly importing seafood, especially high-value products from foreign sources[3]. Marine ecosystems around China are overexploited and degraded, reducing China's ability to produce high-value seafood and rendering its ecosystems more vulnerable to climate change and at risk of collapse. Few fisheries in distant waters offer significant potential for additional supply and often lack the necessary investments for sustainability (e.g. technical expertise, technology, financing). Similarly, overseas sources of mariculture products will need to be developed sustainably in order to ensure a reliable long-term supply. If China is to meet the growing demand for high-value seafood that its increasingly affluent population

1 Bureau of Fisheries (BOF), Ministry of Agriculture. China Fisheries Statistical Yearbook. Beijing:China Agriculture Press, 2017.

2 Bureau of Fisheries (BOF), Ministry of Agriculture. China Fisheries Statistical Yearbook. Beijing:China Agriculture Press, 2017.

3 MOA. 农业部：2017 年我国水产品进出口贸易再创新高预计 2018 全年贸易顺差将收窄 [EB/OL].http://www.moa.gov.cn/xw/bmdt/201803/t20180314_6138388.htm accessed on 2 June, 2018.

prefers, it will need to restore its domestic marine ecosystems, make domestic seafood production sustainable and find ways to encourage other countries to manage their resources sustainably, all while building resilience to climate change. These changes can protect and grow the diverse industries beyond the production sectors that depend on seafood in China.

In 2016, the total value of China's fishery economy was 2.366 trillion yuan, of which the output value of fisheries was 1.2 trillion yuan, and the related industry and construction, circulation and services' output value was about 1.16 trillion yuan. The largest proportion of fishery output value was freshwater aquaculture (581.3 billion yuan), followed by mariculture (314 billion yuan) and marine capture fisheries(197.7 billion yuan). Recreational fisheries accounted for 66.45 billion yuan and continue to rise. There were about 9,700 aquatic product processing enterprises in the country, and the annual processing capacity was about 21.65 million tons, including 17.75 million tons of processing capacity for seafood (Figure 1-2).

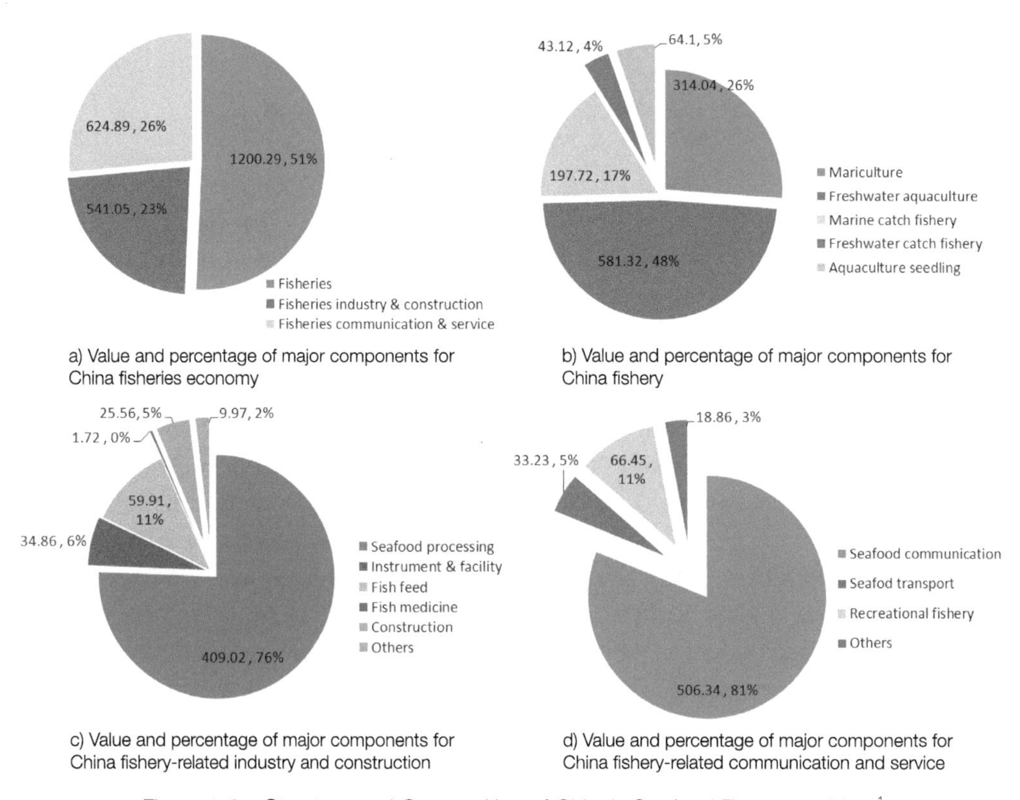

Figure 1-2 Structure and Composition of China's Seafood Economy, 2016[1]

1 Bureau of Fisheries (BOF), Ministry of Agriculture. China Fisheries Statistical Yearbook. Beijing:China Agriculture Press, 2017.

1.2 Status and Trends

1.2.1 Aquatic food demand in China

For nearly half a century, China's aquatic food apparent consumption has been steadily growing, and now represents around 37% of all aquatic food produced globally. In the last decade, 65% of the increase in the global demand for aquatic food can be attributed to China[1]. From 1993—2013, growth in per capita apparent consumption of aquatic food in China increased 5% annually to 38kg/person, and reached 49.9 kg in 2016, already double the global average of just over 20kg/person[2]. The rate of increase in aquatic food production has far outpaced China's relatively flat population growth, so aquatic food is becoming a larger component of the national diet[3]. With a population of over 1.3 billion people continuing to grow modestly at 0.5% annually, the supply of aquatic food needed to meet the demand in China will continue to increase[4]. A recent estimate predicts that, over the next decade, 53% of global aquatic food consumption will increase solely due to demand in China, with per capita consumption expected to reach 50kg/person by 2026[5]. Economic value associated with this demand could increase at a 4% compound annual growth rate until 2020, reaching a market value of $80 billion by 2021[6].

However, unlike some nations with high rates of aquatic food consumption, China's high rate of consumption does not necessarily reflect strong dependence on aquatic food for nutrition or food security. In 2011, fish comprised only 2% of the diet of China's rural population and 5% of the diet of the urban population, whereas the percentage of land-based sources of protein was nearly three times higher in either population[7]. This rural/urban divide suggests that increasing aquatic food

1 NIKOLEK G, DE JONG B, PAN C. China's changing tides: shifting consumption and trade position of Chinese seafood[EB/OL]. Rabobank RaboResearch Food & Agribusiness Report,2018.https://research.rabobank.com/far/en/sectors/animal-protein/chinas_changing_tides.html
2 FAO. The State of World Fisheries and Aquaculture 2016. Contributing to food security and nutrition for all. Rome. 2016.
3 FAO. The State of World Fisheries and Aquaculture 2016. Contributing to food security and nutrition for all. Rome. 2016.
4 THE WORLD BANK, DATA BANK. Population growth (annual %), China. 2018. https://data.worldbank.org/indicator/SP.POP.GROW
5 NIKOLEK G, DE JONG B, PAN C. China's changing tides: shifting consumption and trade position of Chinese seafood[EB/OL]. Rabobank RaboResearch Food & Agribusiness Report,2018.https://research.rabobank.com/far/en/sectors/animal-protein/chinas_changing_tides.html
6 AGRICULTURE AND AGRI-FOOD CANADA (AAFC). Sector trend analysis: fish and seafood trends in China. Global Analysis Report,2017.http://www.agr.gc.ca/resources/prod/Internet-Internet/MISB-DGSIM/ATS-SEA/PDF/6869-eng.pdf
7 LIU G. Food losses and food waste in China: a first estimate. OECD Food, Agricultural, and Fisheries Paper #66. Paris, France..2013.

consumption reflects growing affluence in China and preference for, rather than dependence on, higher quality food. Indeed, China's demand for aquatic food and other high-value marine products is growing and China is increasingly importing seafood, especially high-value products, from foreign sources[1]. In 2017, imports rose by 21.7% in volume and 21.03 % in value[2]. Growth of China's middle class, improved product handling, storage and transportation infrastructure; and access to new markets are important drivers of changes in Chinese consumer preferences.

1.2.2 Mariculture production in China

China's aquaculture industry has grown markedly for more than six decades (Figure 1-3), increasing from fewer than 100,000 tons in 1950, to 3.6 million tons in 1985, and then to 51 million tons in 2016, making it the largest aquaculture producer in the world, and accounting for around two-thirds of global production. During this rapid growth, aquaculture has made significant contributions to safeguarding market supply, increasing rural income, improving the export competitiveness of agricultural products, improving people's diets and guaranteeing food security. In 2016 mariculture production was 19.6 million tons, or 56% of total seafood production in China. China currently contributes 60% of global mariculture production and now produces considerably more aquatic products through aquaculture than wild capture fisheries in both marine and freshwater systems[3]. Freshwater products account for around 62% of total aquaculture production in China, but mariculture production is still substantial and continues to grow.

1 MOA. 农业部：2017 年我国水产品进出口贸易再创新高预计 2018 全年贸易顺差将收窄 .2018.http://www.moa.gov.cn/xw/bmdt/201803/t20180314_6138388.htm accessed on 2 June, 2018

2 GODFREY M. (2018). China's seafood imports surged in 2017, while its export growth continues to slow[EB/OL]. Seafood Source. 2018. https://www.seafoodsource.com/news/supply-trade/chinas-seafood-imports-surged-in-2017-while-its-export-growth-continued-to-slow

3 FAO. The State of World Fisheries and Aquaculture 2016. Contributing to food security and nutrition for all. Rome. 2016.

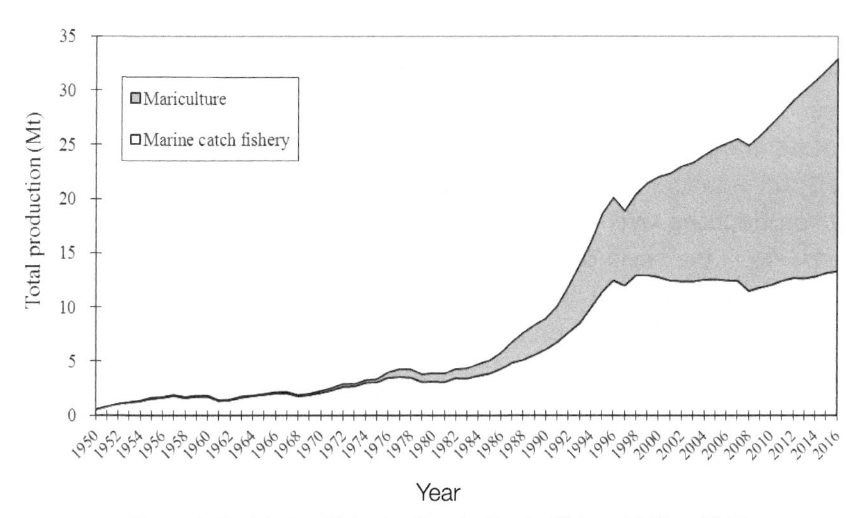

Figure 1-3 Marine Fisheries Production in China, 1950—2016

China's mariculture industry is obviously different from that of other countries in the world. Major aquaculture countries, such as Norway rely mainly on one or a few dominant species, and the mode of operation is also relatively simple; China boasts a diversified range of cultured species, methods and scales. There are more than 70 registered mariculture organisms in China including finfish, shellfish, seaweed and sea cucumbers. A considerable proportion of them are grown by photosynthesis or filter feeding plankton; no feed is needed during the culture process. Only finfish and some shrimps and crabs are fed species; their total production accounts for about 15% of the total mariculture output (Figure 1-4).

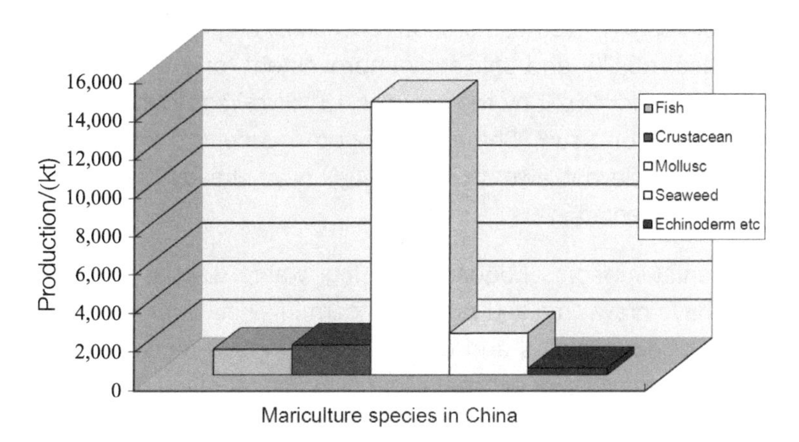

Figure 1-4 Total Production of Mariculture Species in China, 2016[1]

1 Bureau of Fisheries (BOF), Ministry of Agriculture. China Fisheries Statistical Yearbook. Beijing: China Agriculture Press,2017.

Backed by powerful policy, fiscal and technological support, mariculture has been an important driver of the rapid development of the ocean economy in past decades. Since 1950, the industry has enjoyed development opportunities in inland and coastal areas because the central government has adopted a "cultivation-focused" or "aquaculture priority" development policy, resulting in mariculture production doubling every four to five years. Five "mariculture tides", or great leaps forward in the scale of development, have greatly speeded up the scale of production and diversification. Backed by favorable policies, these leaps were further enabled by technical breakthroughs in the culture of kelp, scallops and shrimps; large-scale seedling production of marine fishes; and, more recently, development in mariculture of sea cucumbers and other high-value species. Since 1990, China has made considerable progress in breeding new varieties and strains of aquaculture organisms, disease control, culture technology optimization and harvest mechanization due to technological improvement and financial input from the central and provincial governments. Currently there are thousands of research teams from Chinese universities and institutes working on a full range of aquaculture-related topics.

Since the 12th Five-Year Plan period (2011—2015), China has made a transformation in aquaculture policies by highlighting sustainable development. In April 2015 the China State Council issued the Water Pollution Prevention Action Plan, which clearly stipulates that an ecologically healthy aquaculture will be promoted, with aquaculture prohibition areas drawn for key rivers, lakes and coastal seas; upgrading of aquaculture facilities; strengthening the control over feeds and chemicals; and encouraging deep water aquaculture practices. An upper limit of 2.2 million hectares was also established for mariculture, while "volume reduction & value increase, quality and efficiency improvement, and green development" were set as targets for 2020 by the fisheries'13th Five-Year Plan and more recent regulations. In the long run, China's policy on mariculture will be "health and sustainability", and relevant laws and regulations have already been enacted so as to guarantee implementation.

The environmental impacts of aquaculture (e.g. waste discharge, misapplication of chemicals) have drawn marked public concern in China over the last few years, as awareness of ecosystems and environmental health increases; progress is noticeable in the mass closure of coastal farms and appearance of extensive coastal restoration projects. The clear division of responsibility in the new regulations, as well as the reorganization of ministries and shift in China's administration in 2017—2018, promises to significantly improve ocean and mariculture governance.

1.2.3 Fisheries production in China

China has recognized the importance of fisheries and become more dedicated to their conservation. For example, fishery management in China has evolved considerably in the past half century. Shen and Heino[1] outline four major stages from the mid-20th century to the present. The first stage in the 1950s was characterized by steady economic and technological development, transforming fisheries from a previously underdeveloped and largely artisanal nature to a more significant industrial enterprise. The second stage in the 1960s saw the majority of fishery resources become fully utilized as industrialization commenced. The third stage, spanning the 1970s and 1980s, involved the subsequent predictable collapse of many stocks in the absence of effective management. The fourth and current stage beginning in the 1990s ushered in the era of more rigorous fishery management in China in response to the stock collapses experienced in the third stage.

Despite the rapid rise and expected continued growth of mariculture production in China, the importance of wild capture fisheries is still significant. China still leads the world in marine wild fisheries catch, by far. In 2013, China's fishing fleets harvested nearly 14 million tons, which is nearly three times the catch of the next largest fishing nation, Indonesia[2]. Approximately 90% of the catch is from domestic waters, with the remaining 10% from the distant water fleet (DWF)[3]. While, total catch from China's EEZ has changed little since the mid-1990s, the catch by the DWF has nearly doubled during that time[4, 5]. The national fishing fleet tripled in size from 1990—2010[6], despite domestic catch remaining static. However, in 2017, MOA set the "double control" goal, which stipulated that by 2020, the number of fishing vessels will be reduced by 20,000 and fleet power will be decreased by 1.5 million kilowatts, while also reducing the domestic marine catch to no more than 10 million tons. Therefore, the potential for continued growth in the supply of seafood from foreign waters harvested by the DWF will be an important factor in shaping the future of fisheries and fishery policies in China.

1 SHEN G, HEINO M. An overview of marine fisheries management in China[J/OL]. Marine Policy,2014, 44: 265–272. doi:10.1016/j.marpol.2013.09.012
2 ZHANG H Z. China's fishing industry: current status, government policies, and future prospects. China as a "Maritime Power" Conference, 2015. https://www.cna.org/cna_files/pdf/China-Fishing-Industry.pdf
3 ZHANG H Z. China's fishing industry: current status, government policies, and future prospects. China as a "Maritime Power" Conference, 2015. https://www.cna.org/cna_files/pdf/China-Fishing-Industry.pdf
4 SHEN G, HEINO M. An overview of marine fisheries management in China[J/OL]. Marine Policy,2014, 44: 265–272. doi:10.1016/j.marpol.2013.09.012
5 ZHANG H Z. China's fishing industry: current status, government policies, and future prospects. China as a "Maritime Power" Conference, 2015. https://www.cna.org/cna_files/pdf/China-Fishing-Industry.pdf
6 SHEN G, HEINO M. An overview of marine fisheries management in China[J/OL]. Marine Policy,2014, 44: 265–272. doi:10.1016/j.marpol.2013.09.012

Fisheries remain an important economic driver in China. In terms of global markets, China is among the top three nations worldwide in the value of both seafood exports (#1 at nearly $20 billion) and imports (#3 at $8 billion)[1]. Although mariculture production now far exceeds that of wild fisheries, China's position in the global seafood supply chain, its world-leading volume of wild fisheries catch, and the size of its fisheries labor force mean that management of wild fisheries remains a critical policy issue for social, economic and environmental reasons.

The national fishing fleet in China is incredibly diverse, fishing from coastal waters, across the continental shelf and beyond. Furthermore, the fleet is distributed among 11 coastal provinces spanning China's 18,000 kilometers of coastline and three large marine ecosystems, and catches more than 1,000 species commercially[2]. Generalizing such a vast and dispersed fleet that utilizes a diverse array of resources is not straightforward, although a few traits are widespread. These fisheries for the most part are highly multispecies and use fairly unselective gears; targeted single species fisheries are very rare. New national fishery policies in China intend to address these widespread traits, while also enabling management to be tailored to unique attributes at the regional and local levels.

1.2.4 Coastal and marine ecosystem health in China

China's diverse coastal and marine ecosystems, including estuaries, wetlands, mangroves, coral reefs, seagrass beds, upwelling systems and more, have the potential to provide a basis for China's transition to a blue economy, especially if China's fisheries and mariculture can be managed sustainably. Coastal and marine habitats in China are home to more than 20,000 species, including 3,000 species of fishes alone. China is home to approximately 5.8 million hectares of coastal wetlands, accounting for around 11% of the country's total wetland area. These wetlands provide $200 billion worth of ecosystem services each year, such as food production and shoreline stabilization, accounting for 16% of the total ecological services provided by all ecosystems in the country[3]. Wetlands are particularly important as feeding, spawning, nursery and overwintering habitats for wild fishes and invertebrates. In 2011, China's coastal wetlands provided 28 million tons of farmed and wild-caught seafood, accounting for 20% of the global total seafood

1 ZHANG H Z. China's fishing industry: current status, government policies, and future prospects. China as a "Maritime Power" Conference, 2015. https://www.cna.org/cna_files/pdf/China-Fishing-Industry.pdf
2 LIANG C, XIAN W, PAULY D. (2018). Impacts of ocean warming on China's fisheries catches: an application of "Mean Temperature of the Catch" concept[J/OL]. Frontiers in Marine Science,2018,5. doi:10.3389/fmars.2018.00026
3 MA Z, MELVILLE D S, LIU J, et al. Rethinking China's new great wall[J]. Science, 2014,346(6212):912-914.

production from fisheries and mariculture[1].

Despite their importance, China has cumulatively lost more than 50% of its coastal wetlands,57% of mangroves and 80% of coral reefs, since the 1950s[2,3]. Coastal wetlands continue to disappear at rates around 2.4 times higher than those of wetlands further inland. During the last two decades, a nearly 11,000 kilometer seawall, lining around 60% of China's total coastline and exceeding the length of the Great Wall, has been constructed to defend storm surge and enclose coastal wetlands for mariculture, agriculture and industrial uses[4]. The cumulative reclaimed area rose from an estimated 800,000 hectares in 1990 to over 1.5 million hectares in 2015, with roughly one-third of the total area devoted to mariculture development (Figure 1-5). Rich in biodiversity, the Yellow Sea region contains important fishing grounds and is an important stopover for migratory birds. Since the early 1980s the region has lost 35% of intertidal habitat area due to reclamation, especially along the coasts of Jiangsu and Shanghai at the southern extent. Much of the intertidal zone, especially in the Bohai Sea, is now occupied by mariculture ponds and cages. Loss of habitats can lead to degradation of associated ecosystem functions and services, and ultimately increase the risk of red and green tide outbreaks and vulnerability to natural disasters such as floods and storm damage. It is estimated that the annual economic cost of the loss of coastal wetlands in China is on the order of $46 billion[5].

1 MA Z, MELVILLE D S, LIU J, et al. Rethinking China's new great wall[J]. Science, 2014,346(6212):912-914.
2 BLOMEYER R, SANZ A, STOBBERUP K,et al. The role of China in world fisheries. Directorate General for Internal Policies, Policy Department B: Structural and Cohesion Policies – Fisheries. For the European Parliament's Committee on Fisheries. Brussels.2012.
3 MA Z, MELVILLE D S, LIU J, et al. Rethinking China's new great wall[J]. Science, 2014,346(6212):912-914.
4 MA Z, MELVILLE D S, LIU J, et al. Rethinking China's new great wall[J]. Science, 2014,346(6212):912-914.
5 AN S, LI H, GUAN B,et al. China's natural wetlands: past problems, current status, and future challenges[J]. Ambio,2007, 34:335-342.

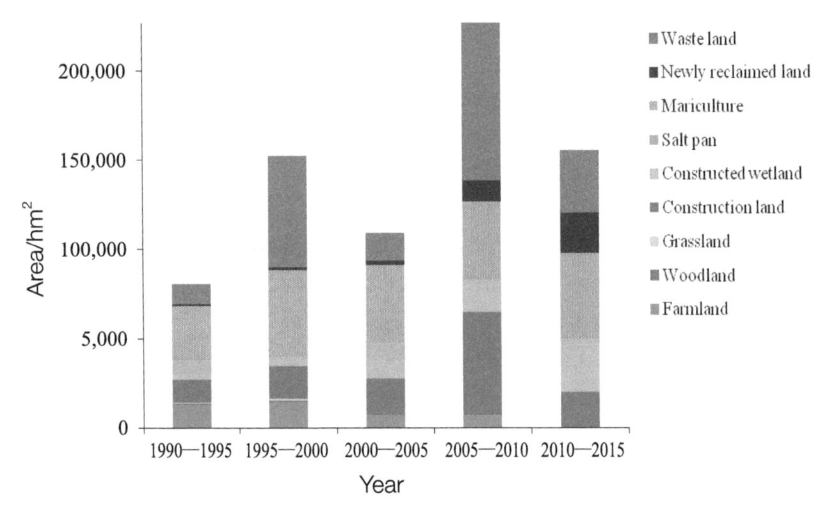

Figure 1-5 Coastal Reclaimed Land Use in China, 1990—2015[1]

It is not only coastal engineering and development that have driven habitat loss in China. Pollutants from mariculture, agriculture and other land-based industries have further eroded key habitats, including those further offshore that are buffered to some degree from alteration of the coastal zone. China's polluted ocean area exceeds about half of its total ocean area, resulting in an estimated economic loss of more than $500 million annually to the country's marine fisheries[2]. Some marine ecosystems, most notably the Bohai Sea and northern Yellow Sea, have been severely degraded and become seasonally hypoxic[3, 4]. Severe eutrophic pollution has occurred in Liaodong Bay, Bohai Bay, Jiaozhou Bay, Yangtze River Estuary, Hangzhou Bay, Minjiang Estuary and Pearl River Estuary, compromising survival of fishes and other living marine resources.

The wild fisheries and mariculture operations that depend upon the healthy marine ecosystems created by marine biodiversity are also one of the major threats to that diversity. Overharvesting and destructive fishing by trawls, fine nets and traps can lead to loss of nursery habitats for many economically important fish species. Over-development of large-scale mariculture can also strain coastal and marine ecosystems and reduce wild fish abundance and benthic biodiversity.

1 Cao L, Chen Y, Dong S, et al. Opportunity for marine fisheries reform in China[J/OL]. Proceedings of the National Academy of Sciences, 2017,114(3), 435-442. https://doi.org/10.1073/pnas.1616583114

2 CCICED Task Force .Ecosystem issues and policy options addressing the sustainable development of China's ocean and coasts[M]. China Environmental Science Press, Beijing, 2013.

3 GAO Y, FU J, ZENG L, et al. Occurrence and fate of perfluoroalkyl substances in marine sediments from the Chinese Bohai Sea, Yellow Sea, and East China Sea[J/OL]. Environmental Pollution, 2014,194, 60-68. doi:10.1016/j.envpol.2014.07.018.

4 ZHAI W D. Exploring seasonal acidification in the Yellow Sea[J/OL]. Science China Earth Sciences,2018, 61. https://doi.org/10.1007/s11430-017-9151-4.

1.3 Challenges to the Management of Living Marine Resources

China's government has taken and continues to take great strides to strengthen management of the country's living marine resources. Doing so is not easy for any country, but it is especially challenging in China, whose living marine resources yield more wild fisheries and aquaculture production, employ more fishermen and fish farmers, and are perhaps under greater pressure from pollution and development than those of any other nation. It is unsurprising, therefore, that despite the progress China has made, it still faces challenges in making its living marine resources sustainable. In our review of the status and trends of China's living marine resources a few themes emerge as requiring the continued attention of the government. First, there is a widespread need for monitoring to improve compliance, to enable management to respond to changing circumstances and emergencies, and to improve the scientific understanding of the ecological system upon which production depends. Second, there is a need to integrate planning to optimize the use of marine space by aquaculture and other commercial uses and to identify critical habitats that need to be protected, restored or enhanced. Third, there is a need to make the production of value, rather than volume of catch, the goal of living marine resource management by continuing to apply output-based management and fishing rights, with which China has already begun to experiment. Fourth, there is a need to understand the impacts of climate change on China's living marine resources and develop policy options accordingly. And finally, there is a need to strengthen laws that protect the habitats and ecological services upon which depends the health of China's living marine resources, and therefore the enormous bounty that China produces from the sea.

What follows is an overview of the challenges China faces in managing the two economic sectors that most affect its living marine resources—mariculture and wild-capture fishing—and the challenges of protecting and conserving the habitat and ecological integrity upon which both industries depend. In addition to such internal challenges as lack of sufficient enforcement, we also describe the additional external challenges presented by climate change and gender issues. Lastly, because China is increasingly looking abroad to supplement its domestic supply of wild-caught seafood, we briefly describe the challenges other countries face in managing their living marine resources.

1.3.1 Mariculture

In China, mariculture permits are issued by the city or county level Ocean and Fisheries Bureau, which is under the administration of both the Ministry of Agriculture (MOA, now the Ministry of Agriculture and Rural Affairs) and the State Oceanic Administration (SOA, now an agency within the Ministry of Natural Resources). Due to a variety of reasons, some of the farms are unlicensed. Even when a fish farm is licensed, change of species or expansion of the scale of operation has not been constrained, which has resulted in pollution, environmental deterioration, prevalence of diseases and increasing difficulties in guaranteeing the quality and safety of aquatic products.

1.3.1.1 Ecological impact of mariculture

The ecological impacts of mariculture mainly include habitat encroachment, environmental pollution and displacement of wild living marine resources. The lack of scientific and rational spatial planning for mariculture is the main cause of many of these problems, which are rooted in policy gaps and manifest most significantly in over-capacity of many mariculture waters in China.

China's mariculture industry is mainly managed through the issuance of sea-use certificates and mariculture licenses. In principle, mariculture operations can only be conducted in national or community-owned waters once the two certificates are in place. Although the two licenses clearly define the water space that can be used, there are no restrictions on the density of the culture, species structure and culture layout[1]. Since farmers tend to culture more profitable species or increase stocking density when prices rise, it is very difficult for the local authorities to know exactly what is being cultured and take measures or control the outputs and discharge in a particular region. Thus, the change of culture species or expansion of the scale of operation is not properly monitored, supervised and constrained. Before the 1990s, this permissive type of management played an important role in enabling the development of mariculture. However, with the continuous expansion of mariculture space and the scale of farming, the unrestricted increase in the cultured biomass has led to increased water pollution, reduced environmental quality, frequent disease occurrence and increasing quality and safety incidents of aquatic products. In addition, unlicensed mariculture has become very common due to conflicts in aquaculture sea use, which has led to overcapacity. Because the current law and legislations, and law enforcement in particular, are not strong enough to control

1 LIU H. National aquaculture law and policy: China. [Eds.] Nigel Bankes, Irene Dahl, and David L. VanderZwaag. Aquaculture Law and Policy - Global, Regional and National Perspectives[M]. Edward Elgar Publishing. 2016.

this situation, functional and powerful tools for spatial planning are needed to place China's aquaculture under strict governance.

Mariculture now occupies one-third of the coastal wetland area and 10% of shallow sea area in China[1]. China has also witnessed large-scale sea reclamation for mariculture[2], including cofferdam and earthen ponds, which has altered a large expanse of coastal landforms and degraded coastal wetland ecosystems. Approximately 240,000 hectares of shrimp ponds have been built in the coastal areas of southeastern China during the past 40 years, largely by destruction of mangroves and seagrass beds. According to the recent nationwide marine inspection[3], reclamation and mariculture development in many provinces has been in violation of laws and regulations, but without adequate restriction by the local authorities. For example, the total area of shallow sea and reclaimed wetlands used for mariculture in Hebei Province is around 18,424 hectares, only 27% of which is licensed. In Jiangsu Province, unlicensed mariculture operations took place in 137 sites and covered more than 13,000 hectares, with 9,954 hectares in conflict with Marine Protected Area (MPA) buffer zones and Ecological Red Line protected areas. Between 1989 and 2000, China lost 12,924 hectares of mangrove forests, more than 97% of which was due to the construction of shrimp ponds.Open violation of law and regulations in mariculture practice is prevalent, clearly reflecting weakness in law enforcement and inaction of government authorities.

Overcapacity in mariculture operations not only displaces many natural habitats, but also degrades remaining habitats by pollution. According to the 2014 report by the National Fisheries Environmental Monitoring Network, pollution of some coastal water bodies in the four China Seas remains severe, while the over-limit ratio for inorganic nitrogen, labile phosphate and petroleum in all samples was 72%, 34% and 39%, respectively, and mariculture was identified as a major source of all three pollutants[4]. Furthermore, antibiotic pollution was also increasing, such as in the waters of the Beibu Gulf[5]. Specifically, erythromycin was detected in 100% of samples at concentrations ranging from 1.10 ~ 50.9ng/L, and sulfamethoxazole

1 LIU H, SU J L. Vulnerability of China's nearshore ecosystems under intensive mariculture development[J]. Environmental Science and Pollution Research. 2017,24: 8957-8966.
2 WANG W, LIU H, LI Y,et al. Development and management of land reclamation in China[J/OL]. Ocean & Coastal Management, 2014,102(B), 415-425. https://doi.org/10.1016/j.ocecoaman.2014.03.009
3 State Oceanic Administration. National Marine Inspectorate's Feedback to Jiangsu Province on the Special Inspects on Land Reclamation. [2018-01-16]. http://www.soa.gov.cn/xw/hyyw_90/201801/t20180114_59954.html.
4 Jia X P, Chen J C, Chen H G, et al. Aquaculture environmental assessment and governance//Tang Q, Environment-Friendly Mariculture Development Strategy: New ideas, New tasks, New approaches [M].Beijing: Science Press.2017.
5 ZHENG Q, ZHANG R, WANG Y, et al. Distribution of antibiotics in the Beibu Gulf, China: Impacts of river discharge and aquaculture activities [J]. Marine Environmental Research, 2012,78: 26-33.

pollution was detected in 97% of samples at concentrations up to 10.4ng/L. Most of the mariculture output in China is from extractive species such as seaweed and mollusks (Figure 1-5), yet fed species including finfish and crustaceans are also cultured in large quantities in China. For fish culture, only about 27% ~ 28% of the nitrogen given as feed are redeemed as fish, and more than 70% of nitrogen is released into the environment[1]. These species clearly contribute to marine pollution, especially when they are cultured in both large scale and high density. Nantong City is among the major whiteleg shrimp production sites in China; the scale of shrimp culture has expanded from 6,700 hectares in 2013 to 12,700 hectares in 2017. The rapid spread of small sheds across mudflats and farmland has led to a series of environmental problems such as soil salinization and over-extracting and polluting of shallow groundwater resources. Similar problems are also evident in major aquaculture areas in Hebei, Shandong and other provinces, and the cleaning and treatment of mariculture-discharged pollutants has become an important challenge for local governments. Control of aquaculture discharge is one major responsibility of the fisheries authorities at all levels, according to the Fisheries Law of China. However this task has been largely neglected or given way to increasing production; lack of monitoring expertise and waste discharge standards may partially share the blame.

An additional concern about the adverse environmental impacts of mariculture is the high volume of feed needed to produce certain fed species (such as fish and crustaceans), often derived from wild capture fisheries harvesting ecologically important forage fish. China's aquatic feed industry has developed rapidly for more than 30 years. Its production increased nearly 24-fold from less than one million tons in 1991 to more than 18 million tons in 2012, representing 41% of global production. Accompanying this growth has been the development of the world's largest aquatic feed manufacturing enterprises. The processing techniques and quality of some aquatic feeds have improved through time. For instance, the feed coefficient of prawn has dropped to 1.0 ~ 1.2, nearly meeting the international standard for efficient production[2]. However, in some cases the feed coefficient, or the ratio of feed inputs to output of mariculture products, is still very high. Excessive use of feeds can exacerbate pollution and habitat impacts caused by other aspects of mariculture operations. Scientific and technological progress is very important for upgrading feed efficiency and comprehensive performance of aquaculture industry.

1 HALL O, HOLBY O, KOLLBERG S,et al. Chemical fluxes and mass balances in a marine fish cage farm. IV. Nitrogen[J]. Marine Ecology Progress Series,1992, 89(1), 81-91.
2 LIU H, SUN L, WANG J, et al. Current status, problems and countermeasures of environmental-friendly mariculture// Tang Q, Environment-Friendly Mariculture Development Strategy: New Ideas, New Tasks, New Approaches [M]. Beijing: Science Press, 2017.

1.3.1.2 Operational and economic factors

Pollution of China's mariculture waters has become increasingly serious, resulting in occasional mass mortality accidents and huge economic losses. On the one hand, the trend of coastal water eutrophication is readily apparent in the increases in occurrence of red tides and green tides; on the other hand, various types of environmental toxins have increased more subtly, but these have caused more and more catastrophic mass mortality of cultured species. According to statistics of the Ministry of Agriculture(now the Ministy of Agriculture and Rural Affairs), in 2014 a total of 284 fishery water pollution accidents occurred across the country, causing direct economic losses of 53.08 million yuan. The loss of fishery resources in 2014 due to the long-term cumulative impact of pollution on fishery habitat was 8.18 billion yuan, of which 6.98 billion yuan accrued in the marine sector, while 1.2 billion yuan was lost in inland waters[1]. These negative external factors pose a severe challenge to the mariculture industry, which relies on science-based site selection and inlet water treatment to operate successfully. At the same time, an improved environmental monitoring and supervision system may also protect the industry from disastrous events.

In 2017, the Zhangzidao (formerly Zoneco Group Co.) sea ranch experienced a significant decline in shellfish production resulting in a 500 ~ 700 million yuan loss in profits. Though several environmental factors were proposed as the cause, so far there is still no agreed upon explanation. The Zhangzidao incident exposed the serious potential problem of sea ranches that China has invested in heavily to construct at a large scale. At present, more than 200 sea ranches have been built or are under construction, yet most of their planning and site selection lack sufficient scientific research, and their operations lack consistent monitoring. These sea ranches are built mainly for commercial purposes and economic benefits. Although huge amounts of artificial reefs have been placed in these sea ranches, there have been insufficient systemic scientific studies and risk assessments on the ecological and natural conditions associated with such operations. The Zhangzidao incident is the result of a number of unfavorable factors and warns us about the importance of conducting scientific and technological research, and strengthening monitoring and governance on sea ranches.

Further adding to economic losses, increases in the prices of feed, energy and labor have eroded what was once reliable profitability in the early days of the industry; wide fluctuations in the prices of mariculture products and declining consumer confidence caused by successive aquatic product safety incidents have further

1 LIU H, SUN L, WANG J, et al. Current status, problems and counter measures of environmental-friendly mariculture// Tang Q, Environment-Friendly Mariculture Development Strategy: New Ideas, New Tasks, New Approaches [M]. Beijing: Science Press, 2017.

exacerbated these impacts. In 1998, more than 300,000 people in Shanghai suffered a Hepatitis A infection from eating blood clams, which revealed the heavy pollution in China's waters. However, at that time the government did not address the degraded environment, but simply banned the sale of blood clams. In the 20 years since the "blood clam incident", China has experienced food safety issues not only in seafood but also in vegetables, cereals, meat, milk and eggs, many of which are due to environmental pollution. Over time, the health and economic impacts of these incidents have slowly motivated stronger government action, driven by consumer concern and confidence. Without comprehensive environmental governance, the aquatic food quality and safety problems cannot be solved completely.

1.3.1.3 Technical challenges

China's mariculture management lacks scientific and technological support. On the one hand, there is insufficient scientific and technological research and development to support management; on the other hand, China still lacks an independent scientific and technological consulting system, and management decisions are sometimes made without or in disregard of scientific considerations. Monitoring and scientific data collection are the most important tools for mariculture governance, and are inadequately deployed by fisheries authorities at all levels. Information system construction for fisheries governance is outdated, mariculture licenses may not correlate to each farm and management authorities at all levels have not set up a routine stock reporting system for respective farms. As a result, local Ocean and Fisheries Bureaus usually do not know exactly which species are cultured and how much is produced at respective fish farms. There has been little monitoring of the environmental impact of mariculture, including the amount of discharge by respective farms, partly due to the difficulty in data sharing between the governance and research institutions. This lack of data combined with the lack of data-support decision making has become a hindrance for effective mariculture governance.

Application of research findings is a long-standing problem in China. Channels for the extension of mariculture research findings are not smooth due to inadequate policy consistency among the government, research institutions and the industry, which means researchers for applied sciences are often unmotivated. Many research findings, patents and reports are written but not commercialized to create value and benefit. Despite these barriers, China has a relatively developed mariculture technology popularization system, with 13,463 aquaculture technology popularization stations (ATPS) at all levels, approximately 37,600 people engaged and around 3.7 billion yuan spent on aquaculture technology popularization each year[1]. This popularization system has played an important role in pushing ahead industrial development and promoting new culture techniques and species. However, because some ATPS leadership does not have adequate professional training, some techniques being promoted are not prominently advanced, while others are neither supported by accurate scientific data nor verified by practice. To inspire creativity in mariculture, China should nurture a deeper respect for science and technology and revise its policies at all levels to encourage innovation and vitality.

Environmental and fisheries policy in China is now evolving to confront many of the environmental impacts of mariculture, which at the same time present new challenges to the industry. Since the 18th National Congress of the Communist Party of China in 2012, China has elevated its commitment to ecological civilization. As a result, China's environmental quality has improved, especially as the new National Environmental Protection Law entered into effect in 2015, with environmental supervision, ecological auditing, environmental law-enforcement and punishment for infractions increasing as a result. After decades of unconstrained investment and growth, these policy changes have imposed unprecedented pressure on the mariculture industry, which is now faced with a ban on sewage and sludge discharge and the use of coal-fired boilers. The industry are required to retreat once a conflict is identified with an MPA or the Red Line system, tourism, or certain other sea uses. Mariculture enterprises must decide whether to disband or reform their operations to comply. China's commitment to improving environmental quality and reducing carbon emissions will affect all economic sectors, including mariculture; as this poses an imminent burden on the industry, swift action should be taken to improve technology to enable less-carbon intensive pollution treatment and green development.

1.3.2 Fisheries

Overfishing is not only the most significant problem facing many fisheries worldwide, it is perhaps the most significant issue facing marine ecosystems overall. This is not to say that other issues are not important, and in some locales more severe than overfishing; however, because fishing fleets are vast and widespread, and cause direct mortality to harvested organisms—often with incidental impacts on habitat—overfishing remains a major threat to ocean health at the global scale. Indeed, recent estimates suggest that ending overfishing can promote recovery of marine wildlife, including mammals, birds and turtles, illustrating the severe ecosystem-level

1 Bureau of Fisheries (BOF), Ministry of Agriculture. China Fisheries Statistical Yearbook. Beijing:China Agriculture Press, 2017.

effects of overfishing[1]. The causes of overfishing are diverse and vary from fishery to fishery, though two of the most pervasive drivers include perverse economic incentives that are not aligned with needed environmental outcomes, and fishers being disconnected from decision-making processes and therefore less likely to accept and comply with regulations.

China is no exception to the severe consequences and drivers of overfishing. Marine ecoregions in China were once world famous for rich fishery resources and high quality seafood products. However, overfishing combined with other sources of deterioration of the marine environment in the past four decades has resulted in increasing occurrences of the "rare fish in the sea" phenomenon, whereby species that were once major components of China's fishery yield have become infrequent in both the catch and the ecosystem. Fishing grounds in the Zhoushan archipelago, for example, covering an area of 220,000 square kilometers, were once so productive that the region came to be known as the home of fishing in China and the granary of the East China Sea. However, catch of large yellow croaker in the region that reached 170,000 tons in 1957 dropped to only 400 tons in 2015, a decline of more than 99%.

Declining trends in other Chinese fisheries have been similar, especially for the highest value species, although losses have been offset by increased catch of more abundant and low-value species[2]. New opportunities are also emerging in recreational fisheries and the DWF; although, if recreational fisheries and the DWF are not managed sustainably, these transitions might simply mean exporting overfishing problems in domestic commercial fleets to other fleets. In many cases, a similar combination of factors underlies fishery declines, including unconstrained development of overcapitalized fleets, technical challenges with stock assessment and management, and socioeconomic impacts associated with management reform and resource recovery.

1.3.2.1 Overcapacity

Nationwide catch in China climbed dramatically in the 1980s due to growth in the number of vessels and fishermen that accompanied growing recognition of fisheries as a significant economic driver; technological improvements that enabled fishermen to catch fish despite declining stocks; and government subsidies. However, eventually natural limits in resource productivity will prevent further increases and even lead to declines in the amount of catch in any fishery. This has been the case

---

1 BURGESS M, MCDERMOTT G, OWASHI B, et al. Protecting marine mammals, turtles, and birds by rebuilding global fisheries[J/OL]. Science, 2018,359(6381), 1255-1258. doi: 10.1126/science.aao4248.

2 SZUWALSKI C S, BURGESS M G, COSTELLO C,et al. High fishery catches through trophic cascades in China[J/OL]. Proceedings of the National Academy of Sciences,2016, 114(4), 717-721. doi:10.1073/pnas.1612722114.

in China where the total volume of catch has changed little since the mid-1990s, but the amount of catch attributable to China's distant water fleet has grown.

Fisheries policy and management have changed overtime in China as these became stronger national priorities. Significant changes to China's approach include a series of progressively more restrictive licensing schemes and periodic vessel surveys to better estimate the total fishing fleet, including unlicensed vessels. These measures have helped to make clear the extent of the significant overcapacity problems facing China's fisheries (Figure 1-6). The total number of fishers in China has steadily increased over time, and although the number of vessels has decreased, the total engine power of the fleet has increased. This means that smaller vessels are being decommissioned while a smaller number of larger vessels go into operation, but the larger vessels require sufficiently large crews which leads to an increase in the fishing labor force. Although an increasing number of these are DWF vessels that leave Chinese waters to fish the high seas or other nations' EEZs, the DWF still accounts for a minority of the catch in China, and excessive pressure on domestic waters by overcapitalized fleets remains a pressing challenge.

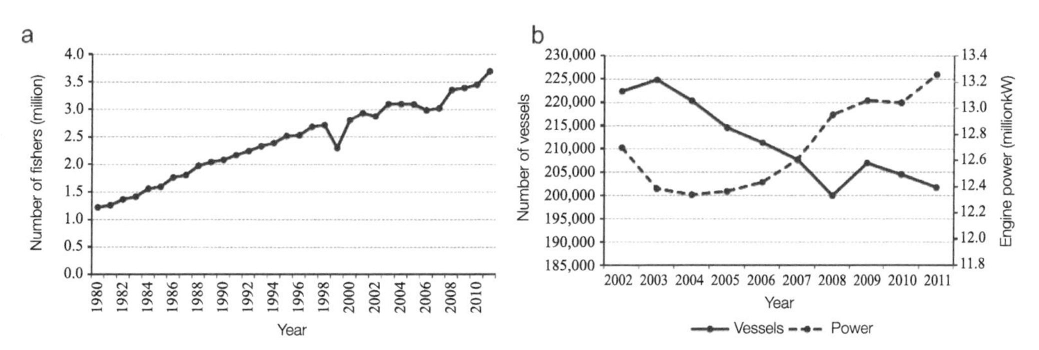

Figure 1-6 Changes in the Total Number of Fishers (a) and Fishing Power in Terms of Both the Number and Engine Power of Vessels (b) in China Through Time[1]

Recent policy changes have aimed to address the severe overcapacity issues in order to bring nationwide fishing power in line with the biological limits of fishery resources. This challenge is made more difficult by the prevalence of illegal fishing gears and unlicensed vessels. For example, in Zhejiang province, the number of unlicensed and illegal fishing vessels peaked at 13,000 in 2013, when total catch in the East China Sea reached three million tons, despite the government having set a target for catch at two million tons. Inadequate numbers of enforcement personnel relative to the large number of fishers and vessels in China make it difficult to

1 SHEN G,HEINO M. An overview of marine fisheries management in China[J/OL]. Marine Policy, 2014,44, 265-272. DOI:10.1016/j.marpol.2013.09.012

adequately monitor the fishing activity and catch of legally permitted vessels, let alone detect and punish illegal vessels.

1.3.2.2 Social and economic factors

Addressing overcapacity in fishing fleets is made more difficult by important and powerful social and economic forces. Pressure to maximize near-term economic gains, especially by impoverished rural communities with few other livelihood options, can cause long-term productivity and stability to be overlooked in favor of increasing yields in the here and now. Similar tradeoffs can be made when an unregulated industry sees an opportunity to capitalize on price trends or establish a strong position in supply chains. These tradeoffs eventually lead to ecological degradation, economic loss and social hardship. Unfortunately, these adverse outcomes encourage fishermen to fish even harder to offset their losses, which causes further degradation, creating a destructive cycle.

Arguably, the solution to fishery recovery and stability is straightforward on the surface: simply reduce fishing pressure to allow stock to recover, and then fish the recovered stock at a conservative rate. With a large fish population, even modest fishing pressure can generate high yields and profits because the underlying resource is abundant and catch efficiency is high. Indeed, a recent global analysis suggests that, despite the high number of overexploited and collapsed stocks worldwide, recovery and sustainable management can lead to even higher yields and profits in the future[1]. However, very real economic and social pressures often prevent governments from undertaking the actions needed to recover. If aggressive action toward recovery is taken, fishing communities could undergo a period of hardship, especially if no other actions are taken to address these communities' economic needs. This may be a short-term problem if stock recovery is rapid, although any period of time without food or income could be devastating to fishing communities with few other viable economic options. Furthermore, overcapacity that contributed to fishery declines in the first place might require some participants exiting the fishery in order to achieve and maintain sustainability going forward.

Subsidies generally compromise the economic and ecological viability of fisheries[2], although time-limited subsidies as investments in keeping fishing communities viable during periods of reform and recover can be an effective strategy. To date, this has not been the nature of fishery subsidies in China. However, in 2015, the Ministry of Finance (MOF) and MOA concluded that the extensive, large-scale and long-term

1 COSTELLO C, OVANDO D, CLAVELLE T,et al. Global fishery prospects under contrasting management regimes[J/OL]. Proceedings of the National Academy of Sciences,2016,113(18), 5125–5129. doi:10.1073/pnas.1520420113

2 World Bank, Food and Agriculture Organization of the United Nations. The sunken billions: the economic justification for fisheries reform[R]. Washington, DC: World Bank.2009.

fuel subsidy policy, which had been implemented since 2006, distorted price signals and was at odds with policies to reduce the size of fishing fleets and constrain harvests. Therefore, the central government announced that the fuel subsidy would be reduced to 40% of the 2014 peak by 2019, and to zero by the end of 2020.

Relocation of fishermen to other livelihoods is perhaps the most important in the series of actions through which Zhejiang has worked to address the "rare fish in the sea" problem and restore the East China Sea fishing grounds. Some displaced members of the commercial fleet have been able to remain employed in fisheries through new opportunities in the growing recreational fishing sector, while others have joined the DWF. Also, like other industries in the coastal zone, over time many fisheries in China have employed larger numbers of migrant laborers; for example, it is estimated that 30% of fishermen in Zhejiang are from other provinces. Generally welfare policies only apply to legal residents of a province, which means that relocation programs do not address the social and economic needs of the whole fleet, and welfare programs are often not applicable to middle aged or elderly fishermen who might still need a source of income to support their families.

1.3.2.3 Technical challenges

Limited stock assessments and issues with data quality and availability preclude rigorous evaluation of the full extent of China's overfishing challenges. However, fisheries lacking scientific assessments are more likely to decline and collapse[1], which means that the perceived poor status of many Chinese stocks may likely be accurate. That China's nationwide fishing fleet continued to grow as total catch remained largely unchanged suggests that stocks are in fact dwindling, as it is likely that only through increasing effort and growth of catch attributable to the distant water fleet that catch has been maintained.

Typically, sound science and sound policy reinforce one another: policies can drive demand for robust science, and science can improve the efficacy of policy. For example, mandates to define clear goals, targets and limits, and requirements for risk-averse management, require scientific information and can provide government and academic scientists a clear focus for their work, as do policies that require consideration of appropriate spatial scales in management and awareness of ecosystem factors that affect and are affected by fishery management decisions. The greater dedication to management in China has led to continued strengthening of national policies, but this progress has not kept pace with the magnitude of overcapacity and environmental degradation that have accumulated since China's fishing industry began to substantially develop in the mid-20th century.

1 COSTELLO C, OVANDO D, HILBORN R, et al .Status and solutions for the world's unassessed fisheries[J]. Science, 2012,338(6106), 517-520.

Despite these problems, China has the benefit of a strong scientific community to help meet its fishery management challenges. Academic institutions and research institutes in China are highly regarded across the globe, and fishery researchers in China have impressive publication records. Yet this capacity is not fully utilized due to structural barriers that unnecessarily compartmentalize individuals and institutions. Transparency and accessibility of fishery data are extremely limited, and consequently universities and provincial fishery research institutes often do not have full access to data collected by the larger regional fishery research institutes, which would enable them to do more innovative research and complete more and better stock assessments. Data-sharing can also pave the way for increased collaboration, which can improve the overall quality and quantity of technical analyses supporting fishery management.

Of course, even if these structural barriers are removed, and data begins to flow more freely and productive collaborations grow across institutions, fishery science in China will only be as good as the data feeding into models and analyses. In particular, accurate data on total catch of each stock, as well as higher resolution data such as catch by different gears, in different seasons or areas, or at different life stages of exploited organisms, are important in any fishery management program. Unfortunately, inadequate catch monitoring also means that these data are often of low quality or altogether nonexistent. Improving monitoring is arguably the single most important challenge facing fishery management in China.

The importance of monitoring lies not only in its scientific benefits, but its value for enforcement and shaping fishing behavior. As monitoring improves in a fishery, the range and effectiveness of management strategies that can be adopted also improves. For example, global experience shows that the implementation of total allowable catch (TAC) management, especially with quota allocated to individuals, vessels, communities, cooperatives or other entities, can be a highly effective strategy for fishery management[1]. However, without effective monitoring, these approaches can incentivize unreported discarding of catch at sea, which results in management targets being missed and scientific assessments based on faulty data. In the absence of the monitoring necessary to allow China to adopt output controls, China has instead relied primarily on input controls. Some of these have had important benefits, but they cannot address all of China's management needs and often introduce inefficiencies that exacerbate socioeconomic strains on fishing fleets and communities.

1 COSTELLO C, GAINES S D,LYNHAM J Can catch shares prevent fisheries collapse?[J/OL] Science, 2008,321(5896), 1678-1681. doi:10.1126/science.1159478

The catch of fisheries in China is highly multispecies due to the nature of China's ecosystem, the use of unselective gears and the absence of limits on the harvest of individual species. Chinese consumers also have a broad palate for seafood so there is less market pressure on fishing fleets to be selective in their catch. Sustainably managing such diversity only increases the need for collaborations across scientific institutions and improved data streams that greater catch monitoring can provide. Furthermore, new scientific tools and management strategies will need to be developed to avoid ecosystem-scale degradation.

1.3.3 Habitat and biodiversity

China has recognized the importance of marine habitats and started to conserve them in a much more dedicated manner. As discussed further in section 1.5, this has been compelled by stronger policy imperatives, backed by government support for increased monitoring, research and protection and restoration projects in many important coastal and ocean areas. Despite this progress, important challenges remain and overall management effectiveness is limited. The most important challenges lie in continued deficiencies in the policy architecture—particularly in the establishment of a strong, clear and comprehensive governance system—and an inadequate technical foundation that precludes both effective implementation of existing policies and creation of stronger policies.

1.3.3.1 Governance issues

The importance of healthy habitats and biodiversity as a basis for sustaining the renewal of living marine resources is still not fully recognized and evaluated in China. There are still significant gaps in the available data with the breadth of living marine resources values such as seafood production, recreation, aesthetics, and regulating services. Despite those gaps, China has introduced a number of laws and regulations to protect coastal habitats and living marine resources, such as the Fisheries Law and Action Plan on Conservation of Living Aquatic Resources. However, implementation and enforcement of these laws and regulations remains deficient. China also lacks comprehensive national legislation for management of living marine resources, which exacerbates conflicts between the mariculture and fisheries sectors, while almost entirely omitting many other values provided by living marine resources from regulatory decisions. Encouragingly, the prominence given to creation of ecological civilization in the 13th Five-Year Plan and the 2018 amendment to the national constitution sets the stage for such a policy, but its development has not commenced. In particular, relevant national policy lacks the following critical requirements:

• Assessment of critical ecosystem functions and status of key living marine resource values;
• Identification of species and areas necessary to protect to maintain those functions and values;
• Requirements to protect or restore those species and areas where possible;
• Monitoring and enforcement to ensure compliance.

The absence of a national living marine resources policy is manifested in a variety of ways throughout the governance system, all of which contribute to habitat and biodiversity remaining at risk. Lacking strong mandates from the central government, local governments lack incentives to prioritize protection and restoration initiatives, and instead favor near-term economic gains. Even where government motivation for conservation is strong, unclear or conflicting authorities among agencies lead to poor or ineffective decision making. One particularly poignant example of this ambiguity is that boundaries for nature reserves and other types of protected areas are often not clearly delineated, so even these bedrock conservation measures are not often effectively implemented. Insufficient enforcement also compromises conservation and restoration efforts.

Furthermore, local decision makers do not have the necessary training and capacity-building in the complex scientific and policy dimensions of living marine resources management, so leaders on the ground are often not sufficiently equipped to meet the challenges they face. Similarly, public awareness of the threats facing, and full breadth of values provided by living marine resources is limited, which means civil society is not voicing its support for conservation or contributing where possible to monitoring, enforcement or problem solving. Although Non-Governmental Organizations(NGOs) are increasingly promoting the broader values of living marine resources and associated conservation needs, the importance of direction from the central government in China means that a national mandate is still essential.

Quite a number of aquaculture species have been introduced into China over the years, including the popular and high-value species such as turbot and whiteleg shrimp. None of these have been subject to robust biosafety checks, and it is unknown if any introduced aquaculture species will likely result in biological invasion. However, cordgrass *Spartina* spp., which was introduced to China as an aquatic fodder species, did become invasive in many locations.

1.3.3.2 Technical challenges

Although the technical basis for productive and sustainable mariculture and fisheries is certainly complex, that of comprehensive conservation of habitat and biodiversity is even more complex due to the much greater number of impacts

and outcomes considered. Adding to this complexity, many living marine resource values cannot be quantified as readily as the revenue generated by seafood sales. For example, whereas a 10% increase in fishery yield will in most cases results in an approximately 10% increase in revenue, it is not as straightforward to say that a 10% increase in abundance of a popular wildlife species will result in 10% more tourism revenue. Both the impact on and value generated by habitats and biodiversity are characterized by uncertainties, nonlinearities and complex interactions. Overcoming these challenges is hindered by insufficient research and monitoring; researchers have too few incentives to tackle these questions rather than focus on those with more directly quantifiable economic outcomes.

Precautionary policies that establish conservative thresholds for habitat and biodiversity conservation, and information about the extent of adverse impacts on those assets, can address this knowledge gap. However, such approaches may incur undue economic costs on the affected industries, or fail to adequately protect habitat and biodiversity if thresholds are not conservative enough or if management actions are unintentionally misplaced.

1.3.4 Climate change

Climate change has had an increasingly dominant effect on global ocean ecosystems, which are affected by changes in temperature and pH, dissolved oxygen, salinity, current patterns and other factors. Although some marine ecosystems are more affected than others—in particular, coral reefs and other biotic coastal habitats that depend upon a very precise combination of environmental variables—all experience these effects to some degree. Changes in ecosystem affect species that are grown through mariculture operations, harvested by fisheries or provide other living marine resource values.

Climate change affects marine organism productivity and distribution, either directly through environmental changes, or indirectly through effects on important prey, habitats or other ecosystem components[1]. Changes in productivity affect the potential available yield for mariculture and fisheries, as well as the recovery potential of species affected by these or other factors. Changes in distribution can introduce species that might compromise mariculture production and affect which species are available for local fisheries, tourism and other uses.

Climate change represents a system-scale impact that will affect management of individual sectors and more comprehensive living marine resource policies; thus,

1 GAINES S, COSTELLO C, OWASHI B, et al. Improved fisheries management could offset many negative effects of climate change[J/OL]. Science Advances, 2018, 4(8): eaao1378. doi: 10.1126/sciadv.aao1378

these factors should be considered to ensure that future policies remain effective as environmental conditions evolve.

1.3.5 Gender aspects

Women account for 50% of the workforce in fisheries and aquaculture worldwide when accounting for the secondary industry sectors that include processing, marketing and selling seafood products[1]. Although they play a vital role in the seafood sector, women in many countries around the world face challenges of inequitable access to opportunity, minimal representation and unequal benefits received from their participation. In addition to achieving fundamental human principles of social justice and fairness, there is a growing body of evidence that gender equality can lead to improved household income, productivity and nutritional security[2]. Addressing the issue of gender inequality is a critical component of sustainable development globally, so much so that it is promoted by UN SDG #5: Achieve gender equality and empower all women and girls.

Approximately 14 million people are employed directly or indirectly by fisheries and aquaculture in China[3]. Women account for 20% of the total professional workforce[4], indicating that about 1.6 million women are employed full time in production and post-harvest work[5]. Aquaculture farms and processing factories help close the gender gap by employing more women as temporary workers. According to our survey, larger aquaculture companies in China tend to employ more male than female workers, with male employees comprising more than 70%, and up to 95%, of the fixed workforce; the ratio of male temporary workers is much lower. Female temporary workers usually earn less than males, and considering the technical and labor-intense requirements, small or family-owned fish farms or seafood processing companies in particular tend to employ more female than male temporary workers(Table 1-1).

1 FAO. The State of World Fisheries and Aquaculture 2016[M]. Contributing to food security and nutrition for all. Rome. 2016.

2 HILLENBRAND E, KARIM N, MOHANRAJ P, et al. Measuring gender-transformative change: a review of literature and promising practices[Z]. CARE USA. 2015.

3 Bureau of Fisheries (BOF), Ministry of Agriculture. China Fisheries Statistical Yearbook. Beijing: China Agriculture Press, 2017.

4 WORLD BANK, FOOD AND AGRICULTURE ORGANIZATION AND WORLDFISH CENTER (WB/FAO/WFC). Hidden harvest: the global contribution of capture fisheries, World Bank, Report No. 66469-GLB, Washington, DC: World Bank, 2012: 69.

5 Bureau of Fisheries (BOF), Ministry of Agriculture. China Fisheries Statistical Yearbook[M]. Beijing: China Agriculture Press, 2017.

Table 1-1 Employment (Including Full-Time and Part-Time) in China's Fisheries(FI) and Aquaculture(AQ)

Year			1995	2000	2005	2010	2012	2016
China	FI+AQ	(thousands)	11,429	12,936	12,903	13,992	14,441	13,817
		(index)	89	100	100	108	112	107
	FI	(thousands)	8,759	9,213	8,389	9,013	9,226	9,226
		(index)	104	110	100	107	110	110
	AQ	(thousands)	2,669	3,722	4,514	4,979	5,214	5,022
		(index)	59	82	100	110	116	111

Source: FAO[1]. The index is with respect to the 2005 baseline.

Since the 11th Five-Year Plan, efforts have been directed towards creating new economic and social conditions throughout the rural areas, including in fishing villages[2]. Though women are expected to take on greater economic roles in the coastal economies, there remains a considerable knowledge gap about the role of women in fisheries and aquaculture. Additional efforts are needed to understand the gender gap and improve women's education, social and economic opportunities and responsibilities in fisheries and aquaculture in China.

1.3.6 Living marine resources in other countries

Chinese fisheries are increasingly looking to overseas fisheries to help meet seafood demands beyond the scope of domestic production. Under the 13th Five-Year Plan, the contribution of China's distant waster fleet to the country's total wild catch is expected to grow from 14% in 2017 to 23%(Huang Shuolin, personal communication). However, today most of the world's assessed fisheries are fully exploited, meaning that target yields are being obtained and no additional yield is likely. Most of the world's assessed fisheries are fully exploited, meaning that target yields are being obtained but no additional yield is likely. Around one-third of assessed fisheries are overexploited or collapsed, which means that current yields are either unsustainable or have already declined(Figure 1-7). Those fisheries that have collapsed represent a potential source of additional yield, if stocks can be rebuilt and renewed harvests are at sustainable levels[3]. The proportion of overexploited and collapsed fisheries might actually be higher given that many countries do not have the capacity to assess their fisheries, and unassessed

1 FAO. The State of World Fisheries and Aquaculture 2014. Rome. 2014.
2 XU S J, YH XU, YL HUANG,et al. Women's roles in the construction of New Fishing Villages in China, as shown from surveys in Zhejiang Province[J]. Asian Fisheries Science, 2012,25S:229-236.
3 COSTELLO C, OVANDO D, CLAVELLE T,et al. Global fishery prospects under contrasting management regimes[J/OL]. Proceedings of the National Academy of Sciences,2016,113(18), 5125–5129. doi:10.1073/pnas.1520420113

fisheries, which are not included among the global trends, are more likely to be depleted[1].

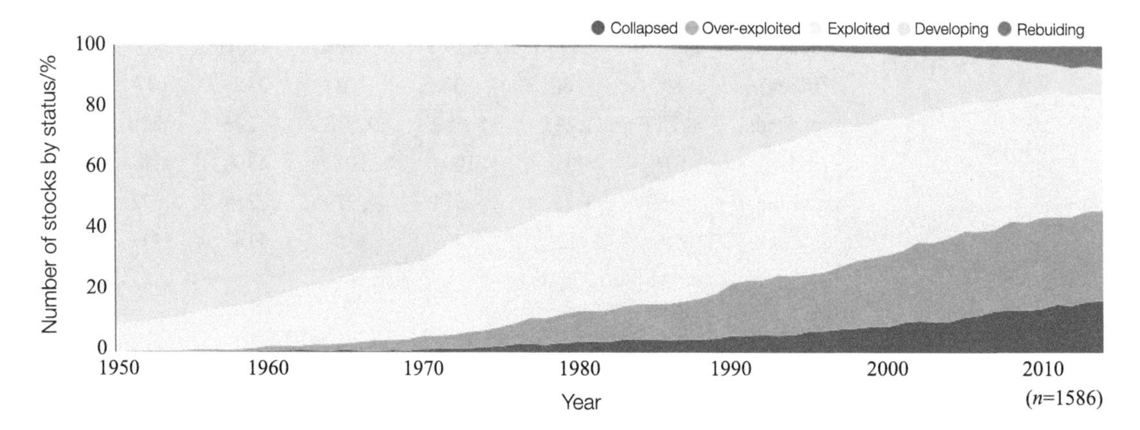

Figure 1-7 Status of World Fish Stocks, 1950—2010[2]

Overfishing is the most significant problem facing many fisheries worldwide today. Because fishing fleets are vast and widespread, and cause direct mortality to harvested organisms, often with incidental impacts on habitat, overfishing remains the major threat to ocean health and biodiversity at the global scale. Indeed, recent estimates suggest that ending overfishing can promote recovery of marine wildlife, including mammals, birds and turtles, illustrating the severe ecosystem-level effects of overfishing[3]. One important driver of overfishing is that many countries, especially in the developing world, lack the scientific expertise, governance capacity and financial resources needed to manage their resources sustainably. Of course, other stressors beyond fishing affect fishery resources, including pollution, coastal development and, perhaps most significantly, climate change.

Encouragingly, bioeconomic modeling suggests that improved fishery management can counterbalance impacts of climate change, which, depending upon the severity of climate change the world experiences, could enable overall worldwide fishery yields to be maintained, or perhaps even modestly increased[4].

While overfishing and the struggles of mariculture management have been globally shared experiences, the solutions and technical expertise accumulated by

1 COSTELLO C, OVANDO D, HILBORN R, et al .Status and solutions for the world's unassessed fisheries[J]. Science, 2012,338(6106), 517–520.

2 PAULY D,ZELLER D. Sea Around Us Concepts, Design and Data.2015. http:// www. seaaroundus.org

3 BURGESS M, MCDERMOTT G, OWASHI B, et al. Protecting marine mammals, turtles, and birds by rebuilding global fisheries[J/OL]. Science, 2018,359(6381), 1255-1258. doi: 10.1126/science.aao4248.

4 GAINES S, COSTELLO C, OWASHI B, et al. Improved fisheries management could offset many negative effects of climate change[J/OL]. Science Advances, 2018, 4(8): eaao1378 DOI: 10.1126/sciadv.aao1378

developed countries in addressing those issues has not been as universal. In fact, research suggests that as developed countries began to regulate their fisheries throughout the 1990s, there was a shift in focus to the largely unassessed fisheries of developing countries[1]. Developing countries' fish production, both wild caught and farmed, has doubled in the last 30 years and now accounts for over half of global fish exports[2]. The fisheries we know the least about and that have the least management in place are being increasingly exploited.

1.4 Progress on Policies for Management of Living Marine Resources in China

Commitment to ecological civilization has taken a firm hold in China's national policies. Through the 13th Five-Year Plan, President Xi Jinping's report to the 19th National Party Congress, the 2018 Constitutional Amendment and other prominent policy instruments, China is repeatedly making a clear declaration that economic progress and social evolution will only continue in ways that ensure environmental protection. Indeed, the philosophy underlying modern policies in China recognizes that achieving economic, social and environmental outcomes does not occur at the expense of one another. Rather, these goals can only be achieved, and more importantly sustained, in concert with one another. Economic progress that degrades environmental resources will undermine the basis for many economic sectors, while declining social conditions will incur welfare and other costs that offset increased profits. Conversely, environmental protections that do not allow economic prosperity will create social costs, fail to gain public support and ultimately fail.

China signaled this new philosophy to the world through its clear commitment to the United Nations SDGs, which likewise aim to achieve the triple bottom line outcomes of economic, social and environmental progress. In 2016, China developed and released a national implementation plan for the 2030 Agenda for Sustainable Management, in which it has created action plans for each SDG target. Since implementation of the plan, China has already undergone one round of reporting on its targets. There have already been notable achievements in environmental protection, for example, in the areas of energy consumption and air pollution, which relate to SDG #13 on combating climate change. In 2016, China's energy

1 WORM B, HILBORN R, BAUM J, et al. Rebuilding global fisheries[J]. Science, 2009,325(5940):578-585.
2 ROHEIM C. Trade liberalization in fish products: impacts on sustainability of international markets and fish resources. In A. Aksoy and J. Beghin, (Eds.), Global Agricultural Trade and Developing Countries. The World Bank.2004.

consumption per unit of GDP fell by 5%, while carbon dioxide emissions per unit of GDP fell by 6.6.%[1]. China has also made tremendous progress in addressing SDG #2, which is focused on poverty alleviation. Disposable income per capita increased by 6.3% in real value, while the number of rural residents living in poverty was reduced by 12.4 million; based on this progress it is estimated that China may be able to achieve its SDG #2 target 10 years ahead of schedule[2].

With regard to SDG #14, China has taken several significant steps to conserve and sustainably use the living marine resources, including drawing redlines for conservation that place 30% of China's sea areas and 35% of coastlines under redline development. It has expanded the total area under protection and intensified law enforcement; imposed more stringent standards for discharging pollutants into the sea; and improved pollution treatment facilities and sewage pipe networks in coastal regions. The government has increased subsidies for reduction of fishing boats and offered subsidies for scrapping fishing vessels, setting limits on the number and total power of fishing boats that decline over time. Finally, China is providing assistance to build aquaculture facilities in countries along the Belt and Road Initiative (BRI)—one of the country's most prominent and powerful vehicles for international engagement in the modern era—thus emphasizing the need to work towards environmental and social goals in tandem with economic goals.

In the maritime realm, the path for achieving a triple bottom line internationally through the BRI has been outlined in the Vision for Maritime Cooperation under the Belt and Road Initiative issued jointly by China's State Oceanic Administration and National Development and Reform Commission in 2017. The Vision addresses the importance of living marine resources and specifies the need to improve management of mariculture and fisheries, and conservation of habitat and biodiversity across the region.

China's international leadership through BRI and other channels will be stronger as it develops domestic success stories that can inform policy changes in other countries. To that end, a series of important domestic policy developments that are improving living marine resource management in China will play an important role. As discussed previously, even before the ecological civilization mandate was elevated during the 13th Five-Year Plan period, China began to address the unreasonable growth and environmental impacts of mariculture through new

1 Ministry of Foreign Affairs of the People's Republic of China (MFA). China's progress report on implementation of the 2030 Agenda for Sustainable Development. 2017.https://www.fmprc.gov.cn/web/ziliao_674904/zt_674979/dnzt_674981/qtzt/2030kcxfzyc_686343/P020170824650025885740.pdf

2 Ministry of Foreign Affairs of the People's Republic of China (MFA). China's progress report on implementation of the 2030 Agenda for Sustainable Development. 2017.https://www.fmprc.gov.cn/web/ziliao_674904/zt_674979/dnzt_674981/qtzt/2030kcxfzyc_686343/P020170824650025885740.pdf

policies, such as the setting of an upper limit for total mariculture area by 2020, and the "volume reduction & value increase, quality and efficiency improvement, and green development" principles put forward by the Fisheries 13th Five-Year Plan. Permitting, monitoring and enforcement are also becoming stronger, which is leading to reduced overdevelopment in the coastal zone, although continued improvements in all of these areas are needed.

The 13th Five-Year Plan prompted MOA to issue a sweeping and ambitious new national fishery policy early in 2017. The policy sets national targets for reductions in total catch and the number and total engine power of fishing vessels. Furthermore, the policy attends to the detailed mechanics of fishery management at provincial and local scales, including improved stock assessments, monitoring to support stock assessments and ensure compliance with regulations, use of fishing rights and opportunities for collaboration in management by the fishing industry, and a shift toward greater use of output controls. Because policy evolution in China relies so strongly on pilot projects that enable provincial and municipal governments to figure out how to operationalize national objectives, and because output controls represent such a fundamental change from how China has managed its fisheries to date, the MOA policy also initiated pilot projects in total allowable catch (TAC) management in five coastal provinces.

The MOA fishery policy focuses not only on improved management of harvested species, but on protection of ecosystems as a whole from adverse effects of fishing. These provisions will be important steps toward more holistic and integrated management of living marine resources. Other policies and pilot projects have also taken important steps in this direction. For example, the integrated ecosystem-based planning pilot in Xiamen has aimed to define space for mariculture, fisheries, shipping and other uses, while minimizing the impacts of these uses on one another and ensuring conservation of critical ecological resources. The pilot strives for this complex balance in a heavily urbanized coastal area, which requires a strong technical foundation and co-management by multiple agencies and levels of government. Notably, the project has been successful in protecting a small but stable population of the Chinese white dolphin, the northernmost extent of the species.

Integrated management of living marine resources might be most effectively achieved not only through holistic policies and coordinated governance, but also by harmonizing management with local traditions and values. Since 2013, China has begun moving in this direction on land through the "Beautiful Countryside" concept, which formally promotes human wellbeing, quality of life and cultural factors into policy. Building from this concept, "Beautiful Fishing Villages" can be a powerful means for incorporating local perspectives, objectives and knowledge into living marine resource management strategies tailored to the unique social, economic and ecological attributes of different places.

1.5 International Experience in Management of Living Marine Resources

Given the scale of China's population, length of its coastline and sheer number of people dependent on living marine resources for their livelihoods, China faces some unique challenges. However, there is still much that can be gained from examining approaches and solutions from other nations, which have shared many of the same fundamental challenges in the management of living marine resources. The following case studies demonstrate ways in which other countries have found opportunities to enhance management of living marine resources in the areas of strengthened monitoring, holistic and integrated spatial planning, managing climate change impact, promoting long-term value over volume in marine fisheries and helping accelerate and scale solutions internationally through exchange. These stories can provide valuable experiences and ideas to draw from as China forges its path towards ecological civilization.

1.5.1 Strengthening monitoring

When living marine resource management is attempted without accurate or complete information, challenges inevitably emerge. In the context of living marine resources, monitoring can involve data collection to improve the performance or management of aquaculture operations, fisheries or ecosystem health. Monitoring can help indicate the success or failure of management, sometimes exposing unexpected patterns or information, which can help advance the state of the science of a particular issue. In addition, monitoring can also help detect non-compliance of rules and regulations, which can lead to stronger enforcement. Therefore, while increased efforts in monitoring can lead to better management, decision making and outcomes, lack of capacity and funding often poses a barrier to expanding such efforts. The following case studies demonstrate the value of monitoring and explore innovative ways to conduct monitoring to advance living marine resource management.

1.5.1.1 Information transparency facilitates governance in Norway

Collecting, analyzing and sharing data on different marine production systems can help operators, fishermen, business, managers and other stakeholders make more informed decisions about living marine resource management. One example of information being collected and used to improve management comes from salmon farming in Norway.

Norway is the largest exporter of aquaculture products in Europe and the sixth largest globally. However, given Norway's geography, 80% of the population lives less than 10 kilometers from the sea, making it crucial to keep all aquaculture clean and well monitored. For salmon farming, the Norwegian Fisheries Directorate has a website that contains not only the location, capacity and operational status of every facility the government licenses, but also the environmental impact assessment results of each. All licenses for salmon farming in Norway are registered with this Geographic Information System(GIS)-based system and each farm is required to submit weekly stock reports to the fisheries authority through the website. Government inspectors make routine checks on farms on a monthly basis. This near real-time sharing of data greatly facilitates mariculture governance and enables scientists to predict the environmental impact of facilities, such as the spreading of sea lice.

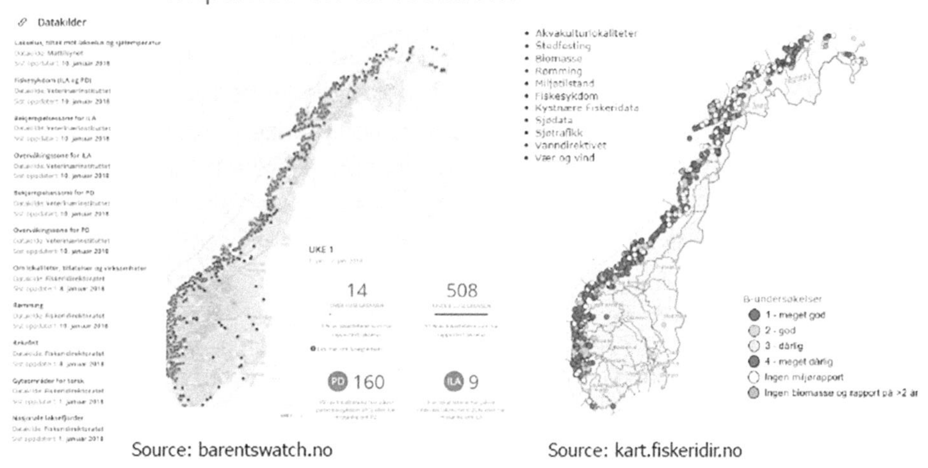

Figure 1-8 Aquaculture Farm Production and Environmental Data Published on the Norwegian Fisheries Directorate Website (Source: Fisheries Directorate Website)

Making data publicly available helps stakeholders understand government decisions, facilitates trust between the groups and creates a strong system of accountability. Scientists can review the data and propose better solutions, and the owners and operators of salmon farms are able to observe nearby environmental issues or disease outbreaks and prepare accordingly. By gathering more data, the government is able to obtain a more accurate picture of the issues affecting and resulting from aquaculture and use that data to improve management overall.

China does not have such toolkits to help with fisheries governance. Although efforts were made by Shandong Provincial Oceanic and Fishery Department, for example, to set up a pilot-scale aquaculture data collecting system (the Fisheries Information Transection Platform), the lack of incentives or pressure on farms to report production data has deemed the system a failure.

1.5.1.2 Electronic monitoring in United States Pacific Whiting fishery

Even when fishers and managers recognize the importance of monitoring and data collection, implementing a system can prove to be expensive and resource intensive. Utilizing technology to assist with the collection and processing of information offers great potential in overcoming such challenges. The following study shows how technology can address this problem by supplementing and even supplanting traditional observer-based monitoring systems in marine fisheries.

Comprised of around 35 vessels, the Pacific Whiting fishery is managed with an annual quota for total allowable catch[1]. Initially, vessels were only accountable for the catch that they brought to shore. However, this incentivized greater offloading at sea of undesirable catch and by catch. Therefore, the government revised the system and created new regulations that required fishermen to account for all catch, including fish discarded at sea. To ensure compliance with this new system, the fishery needed a new comprehensive monitoring system that included at-sea accountability. The nature of the fishery (i.e. many very short trips with very little notice) made it difficult to maintain a human observer system, so instead, the government partnered with Archipelago Marine Research to design an automated electronic monitoring system.

In the electronic monitoring system, boats were equipped with up to four video cameras, fishing gear sensors and a Global Positioning System(GPS) receiver. The data collected by these tools were stored on the boat and retrieved by technicians every two weeks. Fishers were still required to keep logbooks as a means for

1 MCELDERRY H. Electronic monitoring in the shore-side hake fishery 2004—2010. PFMC EM Workshop Agenda (Item B.1). 2013.

comparison with the automated data[1]. Technicians regularly examined the automated data, searching for unreported discards and fishing activity in prohibited areas, and generally collecting vital statistics about the fishery that would prove helpful for fishers and managers. Over the seven years of the program, discard frequency decreased by 90% and data collection success increased to 98%. Unfortunately, the program was brought to an end in 2011 after the government mandated that the fleet had to use 100% observer-based monitoring as part of a broader set of reforms in the region. However, the requirement that a human observer accompany every boat that leaves the harbor has proven costly for the industry, which is now attempting to revert to electronic monitoring once more.

In China, electronic monitoring could help government authorities understand how many fish are being caught in real- or near real-time. Advanced imaging and processing technologies could even help detect which species are being caught, where and at what time, all of which could help the government implement an output-based management system and give scientists a better understanding of the health of China's marine ecosystem.

1.5.2 Holistic understanding and integrated planning

Holistic management of living marine resources will require the need to evaluate and balance tradeoffs between competing activities within the marine space. Research and science underpinning spatial planning, ecosystem services, proper aquaculture density and other activities can greatly improve the information that is used to make decisions about spatial planning. Combined with a robust planning process that involves multiple agencies and stakeholders, this holistic approach to management can help ensure that the benefits from marine uses can be maximized, while minimizing environmental impact and conflict among stakeholders. Below are several examples of how holistic and integrated management have been utilized for living marine resources in other countries.

1.5.2.1 Ecological prioritization in Norway

The ocean area of Norway is six times larger than its land area. In its marine space, a multitude of economic activity takes place, including fisheries, maritime transport, and petroleum and energy. Norway has recently finished designing and implementing a comprehensive system for the coordinated management of all ocean uses. Beginning in 2001, Norway began by implementing a single integrated

1 LOWMAN D M, FISHER R, HOLLIDAY M C, et al. Fishery Monitoring Roadmap.2013.

management plan for one portion of its waters[1]. To create the plan, the government first worked to determine the biological, social and economic basis of the ecosystem and all activities that occur in the space. The government then conducted impact assessments for all use sectors to see how the various activities could interact with or impact each other and the larger ecosystem[2]. This first plan, the Barents Sea-Lofoten area, took four years to develop. The process was then replicated in the remaining two sea sections for a total of three marine plans. Each plan aims to produce greater value in these ecosystems, by pursuing the sustainable use of its resources while simultaneously upholding the integrity of natural ecosystems[3].

To balance these sustainable resource uses with ecological conservation, the planning and management process has involved a tremendous number of government agencies, repeated input from stakeholders and intensive scientific study. One particularly valuable approach to manage such a complex process has been to establish ecological priorities at the start of the process. The main criteria used for prioritization include value—based on productivity or biodiversity—and vulnerability, based on concentration of all organisms, key life stages taking place, abundance of sessile organisms, migratory routes, etc.[4]. Examining these key conditions establishes an ecological foundation which the other uses must work around and not adversely impact. While each marine use is managed individually on a day-to-day basis, management is integrated into a higher level of governance which is bound by these ecological priorities, allowing the system to operate within a broad and connected vision.

1.5.2.2 Ecological restoration in the Chesapeake Bay and South Atlantic region, United States

The Chesapeake Bay, located in the United States' Mid-Atlantic region, is the largest estuary in the country. The ecosystem has been highly altered by heavy land use on all sides (i.e. urban, residential, agricultural), as well as extensive in-water uses (e.g. fishing, boating, shipping, etc.). Over time, this heavy use and high degree of alteration has led to serious deterioration in the water quality and fishing prospects in the bay. Even when stakeholders and governments noticed the problem, the large

1 OTTERSEN G, OLSEN E, MEEREN G I,et al. The Norwegian plan for integrated ecosystem-based management of the marine environment in the Norwegian Sea[J/OL]. Marine Policy, 2011,35(3):389-398. doi:10.1016/j.marpol.2010.10.017

2 PETTERSEN E. Integrated marine management: Norway's methodology and experience [PowerPoint Slides]. Norwegian Environmental Agency. 2015.http://www.varam.gov.lv/in_site/tools/download.php?file=files/text/Finansu_instrumenti/EEZ_2009_2014/7_10_2015_semin_pr/2_Integrated_Marine_Management_Plans_Eirik_Drablos_Pettersen.pdf

3 SCHIVE P. Ecosystem approach- Norwegian marine integrated management plans [PowerPoint Slides]. Norwegian Ministry of Climate and Environment.2018.

4 WINTHER J. Identifying particularly valuable and vulnerable areas [PowerPoint Slides]. Centre for the Ocean and the Arctic & Norwegian Polar Institute.2018.

number of authorities involved on the land and in the water (e.g. private landowners, municipalities, states and the federal government) made it exceptionally difficult to implement and enforce overarching policies.

In response to a grim Congressional research study that identified excessive nutrient loading as the cause of a steep decline in the living marine resources, the Chesapeake Bay Program was formed in 1983. The program is a multi-jurisdictional and multi-stakeholder partnership established to coordinate policies, funding and technical capacity, and set ambitious, quantifiable goals with deadlines. Today, many impacts from unchecked development and utilization remain, but measurable improvements in water quality, habitat and oyster, blue crab and other wildlife populations have been achieved through reducing pollution and other adverse impacts, protecting healthy habitats where possible, and widespread restoration efforts.

This model of integrated habitat protection has also been used with great success elsewhere in the United States. For example, the South Atlantic Fishery Management Council protects essential fish habitat through its Habitat Plan and Fishery Ecosystem Plan; the Albemarle-Pamlico National Estuary Partnership in North Carolina has created the Comprehensive Conservation and Management Plans under the auspices of the National Estuary Program; and the state of North Carolina has developed its own Coastal Habitat Protection Plan. By continuing to consider all uses, while placing a premium on underlying ecology, integrated management plans can help resource users achieve greater collective returns over the long term.

1.5.2.3 Spatial management on the Great Barrier Reef

After a particularly destructive outbreak of crown-of-thorn starfish in the Great Barrier Reef, Australian legislators took action to protect a natural wonder and cultural icon from ongoing environmental issues. In 1975, a federal law created the Great Barrier Reef Marine Park, designed to be multi-use for fishing, recreation and shipping, among other sectors. To design it, the park authority split the park into smaller sections and then zoned within each area[1]. However, throughout the 1990s, more and more scientific evidence accumulated to show that this piecemeal approach was not effective.

Instead of continuing with an ineffectual approach, the government reexamined the evidence and began a new strategy. A comprehensive and science-based planning process was launched to define representative areas covering the range of

1 GERSHMAN S J, BLEI D M.A tutorial on Bayesian nonparametric models. Journal of Mathematical Psychology,2012,56(1):1-12.

habitats, species and communities across the entire park, rather than first breaking the park into sections. It was a long, participatory and consultative process. Authorities considered social and economic impacts and sought alternatives that met environmental goals while minimizing adverse impacts. While the process was lengthy, it was crucial to investigate all of the factors and consider all options, so that the design proved more effective and more durable in the long term. The plan, finalized in 2006, placed 33% of the Great Barrier Reef into no-take zones and protected 70 different bioregions that were defined during the process. Not only has this system proven successful for Australia, which has seen increases in fish density and average size, but internationally the process is a valuable anecdote about the importance of adaptive management and the significance of taking the time at the start of a system to truly learn all of the information and involve all of the stakeholders.

1.5.2.4 Development of marine spatial planning tools

As coastal waters become more crowded with agriculture, fisheries, shipping and a growing list of sectors, managing and planning for each sector separately is proving to be an inefficient and suboptimal approach. Taking on a comprehensive, integrated marine spatial planning process can appear overwhelming, but an increasing number of tools have emerged to assist with the task. SeaSketch, developed by researchers at the University of California–Santa Barbara, is a tool for non-experts to generate hundreds of possible marine spatial plans for their regions. These plans are run through analytics that report predicted biological, social and economic performance; the resulting information can be used to better evaluate benefits and explore tradeoffs among different zoning plans. The platform even allows for stakeholder input on different plans and peer collaboration. The SeaSketch tool has been used in an extensive zoning plan for Barbuda (the Barbuda Blue Halo Initiative) and in the Hauraki Gulf of New Zealand (Sea Change), among other places. Frameworks like SeaSketch continue to emerge and improve, integrating greater amounts of data and introducing more thorough tradeoff analyses[1]. Armed with these tools, managers can more confidently and efficiently make successful integrated plans.

Aquaculture spatial planning is currently being done in pilot-scale in China; for example, Sanggou Bay of Rongcheng City has been divided into aquaculture permitted zone, restricted zone and prohibited zones[2].

1 LESTER S E, STEVENS J M, GENTRY R R, et al. Marine spatial planning makes room for offshore aquaculture in crowded coastal waters[J/OL]. Nature Communications, 2018,9(1). doi:10.1038/s41467-018-03249-1

2 SUN Q W, LIU H, SHANG W T, et al. Spatial planning of aquaculture in Sanggou Bay and surrounding sea areas[J]. Ocean and Coastal Management. 2018.

1.5.3 Managing for climate change

The effects of climate change are already impacting and will increasingly impact all ocean ecosystems and every country's living marine resource management. The threats range from changes in stock distribution and productivity to questions of food security and fisher safety. However, the severity or combination of climate change impacts will differ across contexts. There is no universal solution to this problem. Some stocks will increase in some regions; some will disappear. Some mariculture operations might thrive from temperature changes; some might fail from extreme weather[1]. Countries are already being confronted with these challenges. One contentious example is the so-called "Mackerel Wars" between Britain (backed by European Union) and both Iceland and the Faroe Islands. As Iceland discovered a greater and greater number of mackerel in its waters, driven north by rising sea temperatures, they increased their quota significantly. Without a corresponding decrease in quota by another nation, the mackerel fishery became unsustainable and eventually lost its Marine Stewardship Council(MSC) certification in 2012[2].

While climate change impacts are diverse and a great deal of uncertainty can exist, managers can research and plan for different scenarios. For example, the International Council for the Exploration of the Seas (ICES) is working with NGOs— including Environmental Defense Fund(EDF)— to compile the correct questions and push forward innovative solutions for a changing landscape. It will prove crucial to cultivate strong regional partnerships and develop large ecosystem-based approaches[3]. Institutions should begin to incorporate uncertainty into management frameworks and all new policies should consider climate change from the start.

1.5.4 Promoting long-term value over volume in wild fisheries

International experience shows that when fisheries are sustainably managed, they can provide more food, more prosperous livelihoods and a healthier marine environment. In many examples, fishermen have transitioned from maximizing their catch volume in the short term to successfully increasing the fishery's long-term value. While the management tactics underlying such a transition can vary,

1 DE SILVA S, SOTO D. Climate change and aquaculture: potential impacts, adaptation and mitigation//K Cochrane, C De Young, D Soto, et al, Climate change implications for fisheries and aquaculture: overview of current scientific knowledge. FAO Fisheries and Aquaculture Technical. Food and Agriculture Organization of the United Nations.2009.

2 JENSEN F, FROST H, THØGERSEN T, et al. Game theory and fish wars: The case of the Northeast Atlantic mackerel fishery[J/OL]. Fisheries Research,2015(172): 7-16. doi:10.1016/j.fishres.2015.06.022

3 MOUSTAHFID H. Current actions, identified solutions and opportunities in addressing the effects of climate change on fisheries and aquaculture. Presentation for The effects of climate change on oceans, UN-ICP-18 meeting from 15-19 May 2017.

providing fishermen, communities and the fishing industry with strong access rights has proven effective to flip the incentive equation and cultivate a conservation ethic, as in the below case studies.

1.5.4.1 Territorial Use Rights for Fishing Management in Belize

In 2009, after years of rising fishing effort, but declining catch of lobster and conch, the government of Belize began an effort to improve the country's fisheries governance. More than just important sources of protein, lobster and conch are the country's two most valuable export species and vital for the income of the nation's fishermen. With such high stakes, the government, assisted by NGOs, launched ambitious Territorial Use Rights for Fishing (TURF) pilot projects in two communities. At the pilot sites, access was restricted to local fishermen who had fished in the area historically. These fishermen were then granted the right to fish in exchange for respecting no-take zones and other fishing regulations. This resulted in less competition on the water day-to-day and more investment in the future health of their resources[1]; fishing violations also dropped by 60%.

The government scaled the program nationally in 2016 to include all domestic waters. All conch and lobster fishing is now governed through a network of TURFs and co-managed between the national government and local committees comprised of elected fishers. According to the Health Reefs Initiative, Belize's Southern Barrier Reef has improved in health from 2016—2018 to achieve an overall "good" health rating, one of only three reefs in the Mesoamerican Reef to do so[2]. The initial focus on lobster and conch is now being expanded to the high-diversity finfish fishery through the application of data-limited tools. Fishermen are invested in and supportive of strengthening and expanding the network of marine protected areas, so as to improve the quality of their resources. To leverage the better product quality, the cooperatives have partnered with NGOs to create a local seafood certification program with Belizean restaurants and hotels[3]. This new market of local businesses opens the door to a premium price for a sustainable product and rewards fishers for upholding their sustainable practices in the long term.

1 CASTEÑEDA A, MAAZ J, REQUEÑA N,et al. Managed access in Belize. Proceedings of the 64th Gulf and Caribbean Fisheries Institute. Puerto Morelos, Mexico.2011.

2 MCFIELD M, KRAMER P, ALVAREZ FILIP L, et al. 2018 Report card for the Mesoamerican Reef. Healthy Reefs Initiative.2018. www.healthyreefs.org

3 FUJITA R, et al. Assessing and managing small-scale fisheries in Belize//Salas S, Barragán-Paladines M, Chuenpagdee R. Viability and Sustainability of Small-Scale Fisheries in Latin America and The Caribbean. MARE Publication Series, 2019(19).

1.5.4.2 Individual transferable quota management of red snapper in the Gulf of Mexico

The Gulf of Mexico supplies more than 40% of the United States' domestic seafood. Among the catches of crab, shrimp, groupers and swordfish is the commercially and recreationally prized red snapper. Red snapper underwent a rapid depletion of the population starting in the 1950s. By 1990, spawning potential had declined to just 2%, compared to the target spawning potential of 26%[1]. Recognizing a dire situation, management authorities attempted an aggressive series of reforms, such as creating a catch limit, to improve the situation. Instead, the efforts led to a dangerous race to fish, the repeated lowering of the catch limit and a continuing decline in the population. A new approach was needed; on January 1, 2007, the fishery implemented an individual transferable quota (ITQ) program to reduce overcapacity and eliminate the problems associated with derby fishing.

The ITQ program decreased the number of fishers on the water and assigned each of them an individual share of the fishery's total allowable catch (TAC). The TAC is set annually by the Gulf of Mexico Fishery Management Council based on regular scientific stock assessments and with the advice of a committee of scientific and statistical experts. In 2007, at the start of the program, the TAC was drastically lower than in 2006[2]. In some cases, the harsh reduction and resulting small individual quotas caused real hardship for fishers. However, by 2009, assessments indicated the stock was rebuilding and managers increased the TAC. As the program has continued, the number of fishing trips has decreased, the number of discarded fish has decreased, fish stocks have rebuilt, the catch limit has doubled, and there has been a 100% increase in revenue for fishers (EDF). This revenue increase is due to the fact that fishers are now able to fish when market demand and prices are higher—for example, during the season of Lent—and they are catching higher quality fish. As systems like this have been slowly implemented across the country, there are more and more success stories. In 2017, the number of overfished stocks reached the lowest ever level in the United States[3].

1.5.5 Accelerating and scaling solutions internationally

It is no coincidence that there is such little data about fisheries in developing contexts. These countries' fisheries are often small-scale, multi-species and

1 UNITED STATES NATIONAL OCEANIC AND ATMOSPHERIC ADMINISTRATION (NOAA) (n.d.). History of management of Gulf of Mexico red snapper. https://www.fisheries.noaa.gov/history-management-gulf-mexico-red-snapper

2 AGAR J J, STEPHEN J A, STRELCHECK A,et al. The Gulf of Mexico Red Snapper IFQ Program: The First Five Years[J/OL]. Marine Resource Economics, 2014,29(2), 177-198. doi:10.1086/676825

3 UNITED STATES NATIONAL OCEANIC AND ATMOSPHERIC ADMINISTRATION (NOAA) Status of stocks 2017: annual report to Congress on the status of US fisheries. Washington, DC.2017.

dispersed, making centralized data collection difficult[1]. More fundamentally, there is a severe lack of technical, financial and governance capacity in many of these fisheries. As developed nations innovate and improve their living marine resource management, these lessons should be shared and replicated in developing countries before their living marine resources fall prey to the same misfortune. Domestically, improving the technical, financial and governance capacity of developing country fisheries provides better food security for populations where fish is normally a vital source of protein[2]. Internationally, growing our global knowledge and management of living marine resources ensures more stable markets and healthier oceans for all involved.

The Oceans and Fisheries Partnership is a shared effort between United States Agency for International Development(USAID) and the Southeast Asian Fisheries Development Center to use improved regional relationships to prevent and stop illegal, unreported and unregulated (IUU) fishing and improve the Asia-Pacific region's fisheries and biodiversity. The partnership uses an array of tools grounded in an ecosystems approach to fisheries management. For example, the collaboration pursues the development of catch documentation and traceability systems that help encourage transparency, identify and combat IUU, and improve available data. The partnership also examines entire supply chains to leverage private sector investment to increase impacts, create new market incentives and ensure sustainable sources of funding across all its participating geographies[3]. These large-scale collaborations are essential and benefit all parties by advancing information exchange and scaling knowledge and capacity to address some of the biggest challenges facing fisheries.

1.6 Recommendations

China faces an emerging crisis in its coastal and marine ecosystems wrought by pollution, widespread coastal habitat destruction, overfishing and overcapacity in aquaculture. Furthermore, climate change threatens to not only severely test the resiliency of China's marine ecosystems, but to exacerbate global and regional

1 PURCELL S, POMEROY R. Driving small-scale fisheries in developing countries[J/OL]. Frontiers in Marine Science,2015,2:44. doi: 10.3389/fmars.2015.00044

2 HALL S J, HILBORN R, ANDREW N L,et al. Innovations in capture fisheries are an imperative for nutrition security in the developing world. Proceedings of the National Academy of Sciences, 2013,110(21), 8393–8398. doi:10.1073/pnas.1208067110.

3 UNITED STATES AGENCY FOR INTERNATIONAL DEVELOPMENT (USAID) .USAID Oceans and Fisheries Partnership fact sheet. 2018.https://www.usaid.gov/asia-regional/fact-sheets/usaid-oceans-and-fisheries-partnership

tensions as it puts pressure on aquaculture production and causes wild stocks to shift across borders. China has an opportunity to address these challenges domestically, and play a leadership role on oceans conservation regionally and globally, with a comprehensive suite of actions:

1.6.1 Strengthen legal protections for coastal and marine ecosystems, while promoting sustainable production

China has already begun to transform its sprawling aquaculture industry into a sustainably managed engine of safe, high quality food production, but further progress will require the creation of stronger legal tools. China should consider enacting a new aquaculture law that places limits on facilities' waste discharge and resource use. The law should require science-based spatial limits that take into account the carrying capacity of the local environment, enabling the industry to optimize the value of aquaculture production while minimizing environmental impacts; these limits could be incorporated into China's National Marine Functional Zone plans. The law should also mandate stock reporting by all facilities, authorize routine onsite inspections and include other provisions that mitigate the industry's impacts on coastal and marine ecosystems, such as limits on use of antibiotics or other chemicals.

China has already begun to shift toward limiting the amount of fish caught in its capture fisheries. Such an approach has proven more effective when combined with rights-based approaches that allocate portions of the catch or local fishing areas to the fishing industry and fishing communities. Rights-based approaches will enable the government to address social issues by equitably allocating access to fish among large and small industries, commercial and recreational fishing sectors and small-scale fishing communities. As China improves fisheries monitoring and data, rights-based approaches will become increasingly practical, and fisheries law should be modified to promote their adoption.

Healthy habitats are necessary for productive coastal and marine ecosystems. China should enact a strong Marine Habitat Conservation Law (MHCL) that strengthens protections for coastal and marine habitats and encourages significant new rehabilitation efforts that restore lost ecosystem functions and resiliency. The law should require the development of a network of marine and coastal protected areas large enough to support a biodiverse ecosystem and the production of high-value economic benefits from capture and recreational fisheries over the long term.

1.6.2 Implement a high-tech monitoring system to combat corrupt and illegal activities that undermine compliance and to improve marine science

In China, as in many countries, multiple government agencies struggle to control long coastlines, vast numbers of fishing boats and aquaculture facilities, and marine protected areas and the red line system. Advanced monitoring technologies promise to make this job much easier. China's innovation in sensors, networking technologies and artificial intelligence can help create a transparent system that can operate across agencies, and even globally, to facilitate enforcement and promote compliance with the rules that are established to protect marine ecosystems. Applied domestically, such a system could enable China to expand monitoring to nearly all of its domestic fishing vessels, landing sites, aquaculture facilities and coastal and marine protected areas; applied globally, China could play a leadership role in helping other countries ensure the sustainability of their resources.

In addition to promoting compliance, a high-tech monitoring system has other benefits. It will create a wealth of new data to vastly improve China's understanding of the health of its coastal and marine ecosystems; enable the government to respond in real time to pollution hazards and other emergencies to protect public health and food safety; and it can help other nations understand the impacts of climate change and collaborate with China in designing ways to mitigate them.

1.6.3 Restore lost coastal and marine ecosystem functions needed to support fisheries production, biodiversity conservation and resilience to development, pollution and climate change

China has already taken important steps to conserve coastal and marine ecosystems through the redlining process. However, more should be done to restore lost habitat, including mangroves, seagrass beds, tidal marshes and flats and coral reefs. These habitats provide a valuable array of ecosystem services, including providing nursery and spawning grounds for a diverse array of marine organisms, filtering and detoxifying pollutants, protecting the coast from erosion and buffering the effects of climate change. It is not enough merely to protect existing coastal and sea areas. If China's coastal and marine ecosystems are to withstand the impacts of pollution and climate change and continue to be a source of tremendous prosperity and food production, China should consider the following steps to undertaking a large scale effort, guided by ecological science, to restore lost ecosystem functions and services:

1.6.3.1 Establish a national "marine ecological report card" on the health of China's coastal and marine ecosystems

China's economic goals for marine capture fisheries, aquaculture, marine tourism and other industries should be based on a sound ecological understanding of what the country's coastal and marine ecosystems can support. To develop a strong scientific foundation, China should consider establishing a comprehensive ecological assessment, i.e., a"marine ecological report card," on the health of the country's marine and estuarine zones. The report should assess the cumulative effects of the intense uses of China's living marine resources by fisheries and mariculture, and the impacts of pollution, development, tourism industries and climate change, on the health of China's coastal and marine ecosystems. It should evaluate the integrity of key ecosystem functions and services, such as water and nutrient cycles and critical habitat, and measure their resiliency to climate change and other future pressures. The report should also recommend ways that China can improve the capacity and resilience of coastal and marine ecosystems to support seafood production and tourism and conserve biodiversity. The report should be made publicly available and regularly updated.

1.6.3.2 Develop a national plan of action to restore lost ecosystem functions and services

Because in China, as in most countries, several different government agencies have authority over the country's marine and estuarine zones, China should consider developing a comprehensive plan to coordinate efforts. The plan should seek to ensure that the actions, taken collectively, preserve the value of ecosystem services and functions and optimize China's economic benefits over the long term. The plan should include the Ministries of Agriculture and Rural Affairs, Ecology and Environment, and Natural Resources, as well as the related provincial and local agencies along the coasts. It should also provide guidance for a wide variety of actions; for example, setting catch quotas in capture fisheries, allocating area for mariculture production, protecting habitat through the redlining process, restoring coastal and marine habitat, restricting pollution and designating areas for functional uses through China's marine spatial planning processes.

1.6.4 Create a network of partnerships among countries along the Maritime Silk Road to promote sustainable marine governance and achieve the Sustainable Development Goals

There is but one global ocean and we are all affected by the health of its ecosystems. The Maritime Silk Road Initiative represents a historic opportunity for

China to demonstrate leadership in global marine governance and advance the UN SDGs. Under the Silk Road Initiative, China should consider creating a network of partnerships with countries from Asia, Africa and Europe to encourage mutual learning and promote joint actions that promote a healthy ocean.

China's leadership is important because, while many developing countries face similar challenges in managing their living marine resources, they typically lack China's scientific and technical expertise, governance capacity and financial resources. As a result, the marine resources in these countries are among the most poorly managed and threatened in the world. China can promote sustainability along the Marine Silk Road by promoting information sharing and collaborations that build educational, scientific and technical capacity in partner countries. Key topics could include managing marine resources sustainably, promoting economic development, improving food security for vulnerable peoples, combatting illegal fishing and building the capacity of women in fishing communities and supply chains.

China could also continue to demonstrate leadership by using the Maritime Silk Road Initiative to catalyze the development of regional and global approaches that can mitigate the impact of climate change on living marine resources. In Asia and Africa, climate change could lead to greater conflict among nations as wild fish migrate across borders and changes in the productivity of mariculture and catch fisheries strain the ability of nations to produce food. The Maritime Silk Road Initiative could provide the region with the leadership and institutional platform to develop urgently needed, collaborative solutions to mitigate the impact of climate change on marine ecosystems.

1.6.5 Assess the impacts of climate change on living marine resources and evaluate ways to mitigate the impacts

Climate change is already having a wide variety of significant effects on living marine resources around the world, which threaten to become worse over time. China could develop solutions to this issue by promoting more research into the impact of climate change on the country's capture fisheries and mariculture, and the natural ecosystem services upon which these industries depend. It is also important to evaluate how climate change will likely affect fisheries and food production across the entire Asia Pacific region and around the world. Assessments should take into account the potential impacts of warming waters, acidification, and altered weather, nutrient and water cycles.

In addition, China may wish to consider ways to not only mitigate the effects of climate change, but effectively adapt to it. For example, scientists could examine how provincial governments could work together to manage fisheries, how the industry could remain profitable as species shift in their ranges, how research programs could be designed for breeding heat- or acid-tolerant mariculture species, and how international fish sharing agreements could be developed or strengthened.

Chapter 2

2018 Policy Recommendations to China's State Council Executive Summary

2.1 Introduction: Overcoming Global Shocks While Creating Green Opportunities

The world faces a future likely to be ridden with shocks of many types. Many will involve environmental risks. Important nations seek to withdraw from environmental and other international agreements; trade wars can influence global climate change; biodiversity loss affects poverty reduction. China is already a very significant player in green development. The great challenge now is for China to deepen its domestic progress while at the same time strengthening its international efforts. This is necessary to secure a "Beautiful China" and a Healthy Planet for future generations.

China Council for International Cooperation on Environment and Development(CCICED) Members welcome China's enhanced role in global environmental governance improvements, including its strong commitment to the 2015 Paris Agreement, participation in the UN 2030 SDGs, efforts to green its Belt and Road Initiative (BRI) and its effort on environment and development for South-South cooperation. CCICED Members also believe that it is timely for China to take a stronger role in global ocean sustainability including helping to reduce the plastics and other pollution burdens; and to accelerate efforts under the Global Convention on Biological Diversity (CBD) and other global and regional agreements, especially with a view to Beijing hosting the CDB 15th Conference of the Parties(COP15) in 2020.Clearly the world needs prominent torchbearers to successfully secure innovative paths towards a healthy planet.

2.2 Specific Policy Recommendations

2.2.1 Upgrade China's contribution to global climate governance through enhanced action on climate change mitigation within China

The recent Intergovernmental Panel on Climate Change (IPCC) Special Report on Global Warming of 1.5°C tells us that the gap between action taken since the Paris Agreement and needed global action to avert catastrophic climate change is larger than previously thought. Domestic efforts and international collaboration must be intensified. Innovation will be needed at the systems level—not only technological innovation, but also institutional, policy, economic, business model, consumer and behavioral innovation.

Given China's impressive progress so far, it is evident that China's greenhouse gas emissions can make a try to peak earlier and at a lower overall level. Action on climate change can play a useful role in promoting financial stability, job creation, poverty alleviation, pollution control and health improvement, as well as supply side structural reform.

To capture the opportunity of recent Chinese institutional reforms, the government should:

(1) Institutionalize an effective coordination mechanism, led by the National Leading Group for Climate Change and Energy Conservation and Emission Reduction, for harmonizing action on climate change with multiple strategic goals, based on increasingly ambitious plans in the short term via the 14th Five-Year Plan, in the medium term via a revised Nationally Determined Contribution (NDC) for 2030 and Beautiful China 2035, and in the longer term via 2050 Strategy.

(2) Provide a strong institutional basis for co-management of climate change and air pollution and synergy with other environmental issues across the dimensions of regulation, data transparency, monitoring, enforcement, supervision and accountability. Climate change targets should be incorporated into the existing environmental protection supervision system led by the Central Committee of the Communist Party of China Environmental Protection Supervision Committee. Local capacity building will be essential.

(3)Tighten coal control policies and the promotion of renewable energy and energy efficiency. Specifically, China should end coal quotas and long-term contracts, control industrial coal use and help coal-dependent provinces to transition to other sources of prosperity. A new renewable energy policy framework is needed, and renewable energy subsidies which had already been agreed upon should be fully paid. In terms of efficiency, China is well positioned to lead in the implementation of the Kigali Amendment to the Montreal Protocol by introducing world-leading standards for domestic and exported air-conditioning and demonstrate centralized cooling at scale.

2.2.2 Play a strong leadership role in developing effective post-2020 global biodiversity goals under the Convention on Biological Diversity

COP15 of the CBD will be hosted by China in 2020. This event is a major opportunity to set a new course in global green governance, and a platform to demonstrate China's commitments and achievements towards becoming an ecological civilization and actively participate in biodiversity and ecosystem global governance. The desired outcome would be to dramatically reduce and reverse biodiversity losses in all parts of the world (in short, to "bend the curve" of biodiversity loss). COP 15 is a major opportunity to accomplish objectives noted below.

(1)Establish an effective mechanism to ensure that the CBD strategic goals can be achieved on schedule, including the following considerations: i) active participation of business, civil society and all actors in society; ii) creative implementation mechanisms, periodic review and scaling up instruments; iii) continuously increased ambitions reflected in clearly defined Nationally Determined Contributions (NDCs), as well as contributions by other stakeholders; and iv) actively communicated and aligned CBD goals that promote synergies with other international agendas including climate change and ocean governance, trade and investment and the SDGs.

(2)Showcase China's biodiversity conservation experience to other nations and to the international community. Focus on China's domestic and global efforts, including but not limited to ecological civilization, redlining, green finance, natural resource assets accounting and auditing, strengthened ecological law enforcement, national park-centered nature conservation systems and mainstreaming biodiversity concerns into other sectors. Spatial planning for infrastructure or renewable energy should be done in ways that to avoid, minimize or offset adverse impacts.

(3)COP15 will cast a spotlight on the overseas impacts of China's investment and trade. China should be ready by taking immediate steps to strengthen greening of the Belt and Road Initiative (see Chapter 4 below), and to reduce environmental, climate and biodiversity impacts arising from China's massive imports of commodities such as timber, palm oil, soybeans, and seafood.

(4)Build successful and on-going engagement involving heads of state. There is a need for proactive outreach linked to a proposed Heads of State Summit at the United Nations General Assembly (UNGA) in 2020, and to build a broad movement of support from all actors in society for the significance of the COP15 event similar to what occurred in the Paris Climate Change COP held in 2015. Appoint a Special Envoy for Nature for preparations of COP 15 and beyond.

2.2.3 Develop an ecological civilization approach for China in national and global ocean governance

Marine ecosystems are threatened in many parts of the world's oceans amid unsustainable levels of fishing and marine aquaculture, coastal and offshore habitat destruction, mounting levels of pollution, climate change impacts, and limited efforts on creating marine protected areas and biodiversity conservation. China faces an emerging crisis in its coastal and marine ecosystems wrought by factors such as those mentioned above. Furthermore, through its distant water fleets and seafood imports, China has great influence on marine ecosystems in many parts of the world.

Recommendations related to biological resources:

(1)China should enact a new aquaculture law that emphasizes best practices, and places clear limits and strict enforcement policies on waste discharge. The law should set out science-based carrying capacity limits that can be incorporated into China's National Marine Functional Zoning. The law should mandate stock reporting by all facilities, authorize routine onsite inspections, and include other provisions that mitigate impacts arising from use of antibiotics or other chemicals.

(2)Implement a high-tech monitoring system for marine science assessments to combat corrupt and illegal activities and that will highlight responsible fisheries, habitat and environmental protection.

(3)Develop a national plan of action to restore lost marine ecosystem functions and services. The plan should include actions governed by the Ministries of Agriculture and Rural Affairs, Ecology and Environment, and Natural Resources, as well as coastal provincial and local agencies.

(4)Establish a national "marine ecological report card" on the health of China's coastal and marine ecosystems.

Recommendations related to marine pollution and coastal habitat issues:

(1)Improve real-time monitoring of primary rivers and outlets discharging into the sea. Improve the connection of water quality standards between surface fresh water and seawater. Integrate governance mechanism between the Lake and River Chief System and the Bay Chiefs.

(2)Formulate a national action plan for marine debris pollution prevention and control. Speed up the research and application of innovative approaches for substitution of plastic products and for waste treatment. Recognize the need to mobilize partnerships for action on plastic pollution and invite such platforms to take shape in China. The recently signed accord between China and Canada on reducing plastics affecting the oceans is an excellent example.

(3)Strengthen Chinese research on emerging marine environmental issues of global concern. Priority topics include ocean acidification, ocean plastics and microplastics, oxygen deficiency in hotspot areas, and other emerging marine environment issues of global concern.

2.2.4 Carry out the greening of the Belt and Road Initiative

With its strong emphasis on infrastructure, the BRI requires careful consideration of climate impacts and long-term ecological changes. Environmental impact assessments with public participation should be at its core. To help in the selection and design of projects, there should be alignment of BRI initiatives with the Paris Climate Change Agreement, global biodiversity targets, and the UN 2030 SDGs. China should apply internationally agreed environmental and social safeguards, transparency rules and public participation at an early stage of planning, to reduce environmental and social risks. Several policy recommendations are proposed:

(1)Commission independent feasibility studies and economic, social and environmental impact evaluations as well as solicit the views of the public at local levels. Recruit independent review and verification experts and ensure information transparency. Design and implement a green arbitration mechanism for Belt and Road projects.Realize broad information disclosure based on the Belt and Road Big Data Service Platform on Ecological and Environmental Protection which is publicly accessible.

(2)Enhance the roles of cooperative platforms such as the International Coalition for Green Development on the Belt and Road, launch a "Greening cities along the Belt and Road" initiative and create a network of partnerships among countries along the Maritime Silk Road to promote sustainable marine governance.

(3)Provide support to national environmental agencies along the BRI on human, technical and scientific capacity building. Support the development of platforms for sharing knowledge and experience in green investment such as the Global Green Finance Leadership Program and the Sustainable Banking Network.

(4)Set mandatory requirements for responsible investment overseas (replacing the current voluntary guidelines for responsible overseas investment). Implement gender mainstreaming as part of best practices in BRI projects.

(5)Launch a "Greening the BRI" fund to test and demonstrate the business case for selected sustainable Belt and Road infrastructure investments.

2.2.5 Strengthen green development performance in the Yangtze River Economic Belt (YREB)

The YREB concept is distinctive and represents a significant new way of approaching river basin management for China and indeed possibly the rest of the world:

(1)Strategically focus remediation and restoration efforts on problems with large impact on the overall river basin health. The following actions are needed: i) continue efforts to reduce the volume of solid waste pollution causing serious water pollution in upstream and downstream areas through to the oceans; ii) develop economic incentives for collecting and disposing solid wastes; iii) promote the recycling of waste materials and reduce the incineration rate; iv) improve livestock and poultry farming pollution control measures; v) improve the performance of wastewater treatment plants and treatment of sludge; vi) pay more attention to social concerns through public awareness campaigns on solid waste treatment and recycling activities.

(2)Adopt a multiple stakeholder engagement approach to carefully identify and address any negative impacts on communities and livelihoods. Integrate gender via a multiple stakeholder engagement approach to good governance. Increase public awareness through education campaigns

(3)Develop both compulsory and voluntary instruments that will best ensure robust business-sector participation in conservation finance. "Development offsets" is the most successful example of compulsory approaches internationally. In terms of voluntary approaches, "pay-for-performance" contracts hold significant promise. The government should play a key role in establishing the scientific basis for such contracts.

(4)Some existing government programs (e.g., eco-compensation) should be expanded to include pathways for business sector investments.

2.2.6 Lead green urbanization through technology, planning and policy innovation

With the emerging of the digital and green era, substantial changes of modality and pathway for future urbanization can be envisaged, including content of development, spatial layout, infrastructure, transport and logistics systems, business and organization approach, institutions and policies. Besides technical aspects,

such innovations also involve mindset, theory, development content and approach, organization and business model, institutional mechanism and policy. Therefore, breakthrough and innovation in the following key areas are recommended:

(1)Fully recognize the impact of the digital age and green development on the urbanization mode and avoid using the old mindset for green urbanization planning. Both the market and society should play key roles in determining urban layout and planning.

(2)Fully incorporate green standards into urban-rural planning. Integrated urban and rural development must be considered in the development of green urbanization plan and relevant policies, with comprehensive consideration of impacts on rural economy, ecology, society and culture. Encourage the flow of urban talents into countryside. Gradually open the right of renting and use of rural housing land to urban residents with proper conditionalities.

(3)Promote green technologies that are economically and technically feasible and have major impacts to unleash their potential in energy saving, emission reduction and industrial upgrading. For instance, energy saving technology for indoor air conditioning could be a possible breakthrough.

(4)Give full play to the local spirit of innovation with respect to infrastructure construction, transportation and logistics systems, institutions and policies related to green urbanization. Nature-based solutions to such challenges as intensive storm-water flooding, sea-level rise/storm surges, enhanced urban heat should be evaluated, considered and adopted, as appropriate. Green construction using bamboo is another example.

2.2.7 Find and address synergies among issues

Most of the issues mentioned above are strongly interlinked. Some actions can be taken which contribute to two or more areas of importance. For example, "Nature and Climate Solutions" can achieve both climate and biodiversity goals. Quality reforestation, investments in mangroves and coastal wetlands, and investments to protect watersheds can all be designed to enhance carbon sequestration and optimize biodiversity outcomes, while providing additional ecosystem benefits such as flood protection and soil retention.Reducing overfishing, improving aquaculture management, and restoring coastal and marine habitats will increase seafood economic production value, restore ecosystem functions, and biodiversity. Efforts to reduce climate, biodiversity and ocean impacts of the BRI will strengthen China's position as 2020 CBD COP15 host. By addressing ecological impacts of trade and investment (e.g., on overseas rainforest conversions for beef, palm oil or soybean production), China can and will inspire other countries to take similar initiatives. Synergies that result in new, green livelihoods should be encouraged.

Chapter 3

Progress on Environment and Development Policies in China and Impact of CCICED's Policy Recommendations(2017—2018)

3.1 Foreword

As a high-level policy advisory body approved by the Chinese Government, CCICED is mainly engaged in proposing policy recommendations on major issues of environment and development for reference and adoption by policy makers. At annual general meetings of the CCICED, Chinese and international members have policy discussions based on policy research, generate policy recommendations and then submit them to the State Council of China and central government departments concerned.

CCICED's Chinese and International Chief Advisors Support Group has been commissioned by CCICED Secretariat to draft a report titled "Progress in Environment and Development Policies in China and Impact of CCICED's Policy Recommendations" annually since 2008, so as to better serve policy discussions and recommendations made by Chinese and international members of CCICED in the following ways: i)having a full understanding of major environment and development policies issued by China over the past year; ii)having a knowledge of main policy recommendations proposed by CCICED in the last few years, especially in the previous year, and incorporation of these recommendations into relevant Chinese legislation and policies over the past year. The report contains no assessment of the impact of CCICED. It summarizes and compares China's policy practice and CCICED's policy recommendations, with a view to demonstrating the relevance of CCICED's selection of policy research themes and its recommendations to policy progress, thereby providing reference for CCICED members. This is the eleventh report provided by the Chief Advisors Support Group.

This report sums up the latest regulation and policy progress in China's environment and development since the 2017 annual general meeting of CCICED. With a writing style remaining the same as its predecessors, the report describes CCICED's policy recommendations and corresponding government actions in China in each part and ends with an overview on policy recommendations for readers' reference.

3.2 Preface

The year of 2018 is an extraordinary year as it is the first year of a new era. This year is of great historical significance as it coincides with the 40th anniversary of the reform and opening-up and the 20th anniversary of National Environmental Protection Agency (now Ministry of Ecology and Environment) and is the first year to implement the spirit of the 19th National Congress of the Communist Party of China. Meanwhile, this year is also witnessing major adjustments to and deepened reform of China's government management systems and mechanisms, with unprecedented intensity and depth of reform of central government agencies. In particular, the administration and management system of ecological environment and natural resources in China has been greatly adjusted. Apart from responsibilities shouldered by Ministry of Environmental Protection (MEP), the newly established Ministry of Ecology and Environment (MEE, the successor of MEP) assumes more environmental responsibilities, such as the responsibility of National Development and Reform Commission (NDRC) for dealing with climate change and reducing emissions, the responsibility of Ministry of Natural Resources (MNR) for supervising and preventing groundwater pollution, and responsibilities of Ministry of Water Resources (MWR) for working out water function zoning, setting up and managing drain outlets and protecting watershed environment. In respect of administration of natural resources, the newly established MNR is authorized to work out main functional area planning in place of NDRC, manage urban-rural planning in place of Ministry of Housing and Urban-Rural Development (MOHURD), and administer natural resources such as water, grassland, forest, wetland and ocean in replace of MWR, Ministry of Agriculture and Rural Affairs, National Forestry and Grassland Administration and State Oceanic Administration (SOA), in addition to duties of Ministry of Land and Resources (MLR, the predecessor of MNR). These efforts are far intensive than expected.

China held the 8th National Conference on Ecological Environment Protection this year, which is the highest-profile and exerts a profound influence on China's ecological civilization. The conference witnessed the birth of a major theoretical achievement, namely Xi Jinping's Thought on Ecological Civilization, which becomes the theoretical guide for Beautiful China Construction. This conference has not only strengthened the determination of environmentalists to hold fast to the main battlefield of ecological civilization and keep fighting with environmental pollution and ecological destruction, boosted the confidence in realizing the goal of building a Beautiful China; but also identified building a Beautiful China and developing an ecological civilization as an unshakable national strategic goal, thus opening a new chapter of fully promoting green transition, building a Beautiful China and making ecological progress.

In the past year, Chinese governments at all levels, enterprises and institutions as well as the public, insist on the idea of harmony between human and nature, focus on changes to the main social contradiction in the new era, strengthen ecological environment protection, push forward high-quality development and offer more quality ecological goods to meet people's ever-growing demands for a beautiful life; strive to win the tough battle of pollution prevention and control, one of the three tough battles to build a moderately prosperous society in all respects, and advance the construction of a Beautiful China; accelerate system reform for developing an ecological civilization and building a Beautiful China, promote green development, solve prominent environmental problems, intensify the protection of ecosystems, reform the environment regulatory system, etc. MEE will remain committed to the core goal of winning the tough battle of pollution prevention and control, energetically implement the Three-Year Action Plan on Defending the Blue Sky deeply carry out Action Plans for Air, Water and Soil Pollution Control and deepen the work of pollution prevention and control; carry on strict environmental law enforcement and environmental protection supervision; work with departments concerned to promote air pollution prevention and control in key areas like Beijing-Tianjin-Hebei region; propel ecological progress, adjustment to industrial structure, energy structure, transportation structure and land use structure, and enhance source control; reform and advance the pricing mechanism for green development, vigorously develop green finance, boost the growth of the energy conservation and environmental protection industry; and facilitate positive results in ecological environment protection in the Yangtze River Basin and Xiong'an.

As a policy platform of the Chinese Government and a bond, bridge and window between China and the international community for environmental cooperation, the 6th CCICED is reformed to bring into full play intellectual resources at home and abroad, carry out a lot of innovative and leading policy research in important fields against typical and prominent environment and development problems in China in the new era, with milestones achieved, and have preliminary generated forward-looking, strategic and warning policy recommendations, thereby continuing to contribute to ecological progress in China and sustainable development around the world.

3.3 Xi Jinping's Thought on Ecological Civilization becomes A Guide for Ecological Environment Protection in the New Era

3.3.1 Overview of the National Conference on Ecological Environment Protection

The 8th National Conference on Ecological Environment Protection was held on May 18—19 in Beijing, which was attended and addressed by President Xi Jinping and Premier Li Keqiang, and attended by Wang Yang, Wang Huning and Zhao Leji. Vice Premier Han Zheng made a summary speech.

Apart from speeches made by central leaders, leaders of NDRC, Ministry of Finance (MOF), MEE, Hebei Province, Zhejiang Province and Sichuan Province spoke to exchange their opinions. Present were also members of Political Bureau of the Central Committee of the Communist Party of China, General Secretary of the Secretariat of the Central Committee of the Communist Party of China, leaders of the National People's Congress (NPC) Standing Committee, State Councilors, President of Supreme People's Court, Procurator-General of Supreme People's Procuratorate and leaders of Chinese People's Political Consultative Conference (CPPCC), together with leaders of provinces, cities, districts and municipalities with independent planning status under the national social and economic development, Xinjiang Production and Construction Corps, departments concerned in central and state organs, mass organizations concerned, large state-owned enterprises concerned and military units concerned.

In his address, Xi Jinping stressed that building an ecological civilization was of fundamental importance for the sustainable development of the Chinese nation. Since the 18th National Congress of the Communist Party of China, we have carried out a number of fundamental, groundbreaking and long-term tasks to promote historical, watershed and global changes to ecological environment protection. Developing an ecological civilization in the new era requires compliance with six principles such as harmony between human and nature, that lucid waters and lush mountains are invaluable assets, that a good ecological environment is the most inclusive wellbeing for people, that mountains, waters, forests, farmlands, lakes and grasslands form a life community, conservation of the ecological environment through the strictest system and rule of law, and pursuit of global ecological progress. Efforts should be made to ensure that the quality of the ecological environment will be radically improved and the goal of building a Beautiful

China basically attained by 2035 through accelerated construction of ecological civilization systems. By the mid-21st century, material, political, spiritual, social and ecological civilizations will be upgraded in all respects, green development models and lifestyles formed all-sidedly, harmony between human and nature maintained, national governance system and capacity for ecological environment fully modernized, and a Beautiful China built.

In his speech, Li Keqiang proposed to study carefully, understand and implement the spirit of important speeches delivered by General Secretary Xi Jinping and, guided by Xi Jinping's Thought on Socialism with Chinese Characteristics for a New Era, build ecological civilization systems, strengthen system and rule-of-law construction, remain committed to ecological progress and environmental protection, and strive to win the tough battle of pollution prevention and control.

Han Zheng indicated in the summary speech that efforts should be made to learn carefully and comprehend Xi Jinping's Thought on Ecological Civilization while local governments and departments should focus on the implementation of the thought, introduce specific and practical policies and measures to ensure feasibility and effectiveness.

3.3.2 Main content of the conference

This conference is the highest-profile ever. All members of the NPC Standing Committee were present at the conference except one abroad. Meanwhile, the conference involves rich content and enormous information, and is mainly composed of the following six parts:

Firstly, the conference has raised the significance of developing an ecological civilization to a higher level, namely upgrading it from being of "great importance" for the sustainable development of the Chinese nation as stated in the report to the 19th National Congress of the Communist Party of China to being of "fundamental importance".

Secondly, the conference offers a highly recognition for the achievements in ecological civilization construction since the 18th National Congress of the Communist Party of China with three affirmative words: historical, watershed and global changes resulted. However, it also indicates that China's ecological environment is still very fragile and the results are not stable despite continuously improved quality of the ecological environment and a trend of being stable and improved.

Thirdly, the conference has made three judgments on the status quo of the ecological civilization construction: first, it is in a key period of various pressures and loads; second, it has entered a critical period that requires more quality ecological goods to meet people's ever-growing demands for a beautiful environment; third, it has also come to a period when China is capable of addressing prominent ecological and environmental problems.

Fourthly, the conference has put forward six principles and five requirements for the advancement of ecological civilization construction in a new era. The six principles are harmony between human and nature; that lucid waters and lush mountains are invaluable assets; that a good ecological environment is the most inclusive wellbeing for people; that mountains, waters, forests, farmlands, lakes and grasslands form a life community; conservation of the ecological environment through the strictest system and rule of law; and pursuit of global ecological progress. The five requirements are to accelerate the construction of ecological civilization systems; to promote green development in all respects; to give priority to solving prominent ecological and environmental problems for better people's livelihood; to effectively prevent ecological environment risks; and to raise the level of environmental governance.

Fifthly, the conference emphasizes speeding up the establishment of five sound ecological civilization systems: an eco-culture system based on ecological values; an eco-economic system led by ecology-based industrial development and ecological industrialization; a target responsibility system with improvement of ecological environment quality at its core; an ecological civilization system guaranteed by the modernization of governance system and capacity; and an ecological security system focusing on virtuous cycle of the ecological system and effective prevention and control of environmental risks.

Sixthly, the conference stresses that the Party leadership should be strengthened and that local Party committees at all levels and government leaders should assume primary responsibility for ecological environment protection in respective administrative regions.

3.3.3 Guiding significance of Xi Jinping's Thought on Ecological Civilization

The 8th National Conference on Ecological Environment Protection is of historical significance for speeding up ecological civilization construction in the new era. The core outcome of the conference is Xi Jinping's Thought on Ecological Civilization, which defines the position of "ecological civilization" in national development and reform at a higher level, more broadly and furthest.

As a systematic and complete theoretical system, Xi Jinping's Thought on Ecological Civilization has rich connotations and gives a profound reply to major theoretical and practical questions such as why to build an ecological civilization, what the ecological civilization should be and how to build. It is the fundamental guide for the ecological civilization construction in the new era and the answer to what modern power should be built in the new era.

Xi Jinping's Thought on Ecological Civilization will become the long-term guiding ideology for building a Beautiful China. Moreover, the 8th National Conference on Ecological Environment Protection, which is of historical significance, marks the beginning of efforts to develop an ecological civilization and build a Beautiful China for a new era and definitely will bring the ecological civilization construction in China to a new stage.

3.4 Planning for Environment and Development

3.4.1 Planning based on 2035 planning for environment and development

As CCICED's first recommendation in 2017 states, a 15-year strategy for pollution prevention and control action plan should be developed. The recommendation indicates that clean coal utilization and gas power generation technology, which are being promoted at present, are just temporary choices in green transition, that efforts should be made to make a plan and budget for cancellation of clean coal utilization on a large scale to prevent China from being trapped in the fossil energy-dependent development path, and that adaptation planning should be a key part of the strategic planning.

In line with China's energy development plan, China will actively promote grid parity for and market-oriented development of wind power and photovoltaic power around 2020 whilst ensuring that the proportion of non-fossil energy consumption in primacy energy consumption will reach 15% by 2020. Based on the "two 15-year" overall arrangement set forth in the report to the 19th National Congress of the Communist Party of China, National Energy Administration (NEA) will make development goals and strategies for 2035 and 2050 in advance, begin to develop an outline of energy development strategy on how to achieve the two milestones, study energy guarantee, draw the roadmap and timetable for energy development in the new era and concentrate its efforts on improving energy supply guarantee and people's wellbeing, advancing green development of energy, increasing development quality and efficiency, implementing the innovation-driven strategy, establishing sound systems, mechanisms and legislation, and promoting the building of a community of shared future for mankind. Preliminary ideas are listed below:

In the first stage from 2020 to 2035, China will actively promote market-oriented reform and innovation in systems and mechanisms, and facilitate scale development and technological progress in the renewable energy industry. China will seek to ensure that by 2035, all incremental energy demands at home will be met by clean energy and renewable energy development will come to the stage of incremental replacement.

In the second stage from 2035 to the middle of this century, China will construct a modern energy system led by renewable energy in an all-round way and usher in a stage of stock replacement of fossil energy with renewable energy. By 2050, the proportions of renewable energy consumption in primary energy consumption and power consumption will reach 60% and 80% respectively, and renewable energy will play a key role in energy supply, while China should complete energy transformation before this.

Meanwhile, provinces and municipalities concerned also focus on the year of 2035 in their overall planning. For example, cities like Beijing have developed their urban master planning at the time node of 2035.

Of course, CCICED has also begun the "SPS on Goals and Pathways for Environmental Improvement in 2035" and studies on a final solution by analyzing the environmental improvement goals set forth in the report to the 19th National Congress of the Communist Party of China, analyzing obstacles and drawing experience from international practices.

3.4.2 Regional planning for environment and development released in succession

As CCICED's policy recommendations in 2017 state, decision-making fragmentation should be addressed to build an ecological civilization. Great efforts are required to maintain information sharing between organizations concerned in a certain region and make overall arrangements for utilization of land and water resources, and responsibilities for protecting and restoring the ecosystem should be further defined.

Current practices such as ecological protection red lines and comprehensive planning in the Yangtze River Economic Belt are a successful case of ecological environment protection in the new era.

The Plan for Ecological Environment Protection in the Yangtze River Economic Belt (hereinafter referred to as the Plan) came into effect in the second half of 2017. The Plan requires clear thinking, establishment of hard constraints and improvement, not degradation, of ecological environment at the Yangtze River. By 2020, the ecological environment should be notably improved, the stability of ecological environment fully increased, ecological functions of rivers, lakes and wetland basically restored, and systems and mechanisms for ecological environment protection further optimized. By 2030, main streams and their tributaries should have sufficient ecologic water, the quality of water environment, air should be improved in all aspects, ecosystem services should be significantly enhanced and the ecological environment should be more beautiful. The Plan contains comprehensive planning for the Yangtze River basin by requiring designating the upper limit for utilization of water resources, drawing ecological protection red lines, sticking to the bottom line for environmental quality, promoting environmental pollution control in an all-round way and strengthening efforts to prevent and deal with environmental emergencies.

In order to facilitate the implementation of the Plan, MEE carried out the Waste Cleanup Action 2018, which started from May 9 and ended at the end of June. MEE transferred law enforcement backbones nationwide and set up 150 task forces which inspected and verified dumping of solid wastes in the Yangtze River Economic Belt, urged local governments to rectify problems if any within a definite period of time, punished illegal behaviors if any according to law and disclosed the list of all problems and the progress on rectification, until all rectifications were completed.

After the program began, departments of ecology and environment at all levels filed and investigated environmental illegality problems in question, imposed the strictest punishments on 26 enterprises and seven individuals breaking the law, and cracked down on illegal behaviors through measures such as closedown and attachment, production suspension for rectification and transfer to detention. Currently, fines regarding the 29 problems total more than 2.60 million yuan, and two persons have been transferred to public security departments for administrative detention. For example, with respect to illegal dumping and stockpiling of industrial solid wastes by Hubei Tianmen Yijia Building Materials Co, Ltd., Tianmen Environmental Protection Bureau filed and investigated the case according to law, ordered the company to suspend production and correct, and imposed a fine of 350,000 yuan.

On July 25, 2018, MEE published the rectification of the first batch of problems supervised and handled by ministries in the Waste Cleanup Action 2018, which involved 29 problems regarding illegal dumping and stockpiling of general industrial solid wastes. To break it down, seven problems were found in Hubei Province, four in Guizhou, four in Jiangsu, three in Yunnan, three in Shanghai, two in Sichuan, two in Anhui, one in Zhejiang, one in Hunan, one in Jiangxi and the rest one in Chongqing. According to on-the-spot verification, all these problems have been rectified as required by MEE.

3.4.3 Development planning for green cities and towns

As CCICED's third recommendation in 2017 states, comprehensive reform of ecological civilization and green development should be carried forward. The recommendation indicates that in the course of rapid urbanization, the principle of green development must be observed and efforts should be made to consider integrated and balanced development of urban and rural areas, carry out necessary planning for ecologically sensitive areas and increase the efficiency of utilizing resources such as space, soil and wastes.

The Outline of the Planning for Xiong'an New Area of Hebei Province was officially released in April 2018, which sets forth the construction of Xiong'an New area characterized by high-level green development through measures such as building a scientific and rational spatial layout, creating a beautiful natural and ecological environment, improving the industrial layout and developing a green smart new city.

The planning for Xiong'an New Area attaches great importance to ecological conservation and construction of an ecological security pattern for the new area. Firstly, it contains the planning and construction of "a lake, three belts, nine patches and multiple corridors" to build an ecocity that is integrated into forests which are surrounded by water. "A lake" means environmental governance and ecological restoration at Baiyang Lake, so as to resume its function as "the kidney of North China"; "three belts" mean the construction of circum-Baiyang Lake green belt, circum-initial area green belt and circum-Xiong'an New Area green belt to optimize the ecological spatial structure between the city and the Lake, between urban groups and between the New Area and surrounding areas; "nine patches" mean the construction of nine large forest patches between urban groups and in key ecological conservation areas to enhance carbon sink capacity and biodiversity protection; "multiple corridors" mean the construction of multiple green ecological corridors along main rivers and on both sides of traffic arteries in the New Area, to protect the blue sky, increase the greening rate and promote ventilation and dust precipitation. Secondly, it puts forward large-scale afforestation. Efforts should be made to highlight indigenous tree species and local characteristics through near-nature greening and multiple mixed modes, plant ecological shelter forests and landscape ecological forests in green belts and ecological corridors to form a plain forest network system, thus maintaining connectivity of ecological space. Land greening should be carried out on a large scale to increase the forest coverage in the New Area from 11% at present to 40%. A high-quality ecological environment should be created for the New Area. Urban ventilation corridors and a city-Lake local air microcirculation system should be built to carry fresh air over Baiyang Lake to city center. A pleasant park system for the convenience of citizens should be established, which should be composed of large suburban eco-parks, large comprehensive parks and community parks, so as to achieve the city surrounded by forests and embracing wetland, maintaining forest, forest belt and park within 3km, 1km and 300m reach respectively, with all streets tree-lined and a green coverage ratio of 50%. Thirdly, it proposes to increase the regional ecological security guarantee. An ecological corridor that joins "Taihang Mountain Chain—Bohai Bay" to "South Beijing Ecological Green Wedge—Juma River—Baiyang Lake" should be built to form a regional ecological security pattern featuring connection of mountains and waters and fusion of the south and the north. Efforts should be made to implement conservation and restoration of key ecosystems, optimize the ecological security barrier system and increase the quality of ecosystems.

3.4.4 Seeking to win the tough battle of pollution prevention and control

As CCICED's policy recommendations in 2017 state, as the Air Pollution Prevention and Control Action Plan has first come to a new stage, China should formulate a comprehensive long-term strategic plan covering water, air, soil and marine pollution for the next ten to fifteen years. Deployment of this overall strategy should be completed by 2020, so that China can basically realize the goal of socialist modernization by 2035.

In mid-June, the CPC Central Committee and the State Council released the Opinions on Comprehensively Strengthening Ecological Environment Protection and Resolutely Fighting for Pollution Prevention and Control (hereinafter referred to as the Opinions). The Opinions sets forth that by 2020, the quality of ecological environment will be generally improved, total amount of key pollutants discharged or emitted sharply decreased, environmental risks effectively controlled and the level of ecological environment protection aligned to the goal of building a moderately prosperous society in all respects. It also states that China should ensure, by accelerating the construction of an ecological civilization system, the formation of the spatial pattern, industrial structure, mode of production and lifestyle for resources conservation and ecological environment protection, fundamental improvement of ecological environment quality and basic realization of the goal of building a Beautiful China. By the middle of this century, ecological civilization will be improved in all respects and modernization of the national governance system and capacity for ecological environment will be accomplished.

The Opinions puts forward three requirements regarding the formation of green production models and lifestyles and presents some specific quantitative indexes. For example, in promoting green, low-carbon and circular economic development, the Opinions particularly requires accelerating removal and reform of heavily polluting enterprises and those producing hazardous chemicals in urban built-up areas and key basins, for which city governments concerned should make special plans and make them public by the end of 2018. In promoting conservation of energy and resources, it requires implementing national water conservation actions, improving the water pricing mechanism, advancing the construction of a water-saving society and water-saving cities and controlling the total water consumption nationwide within 670 billion cubic meters by 2020; actively coping with climate change, taking effective measures to ensure the attainment of the action target for controlling greenhouse gas emissions by 2020, etc.

The State Council promulgated the Three-Year Action Plan on Defending the Blue Sky (hereinafter referred to as the Plan) on June 27. The Plan sets forth, after three years of efforts, total key air pollutant emissions should be sharply decreased, together with greenhouse gas emissions, the concentration of fine particular matters ($PM_{2.5}$) further obviously lowered, the number of heavy pollution days remarkably reduced, the quality of ambient air significantly improved and people's blue sky wellbeing substantially enhanced. By 2020, total sulfur dioxide and nitrogen oxide emissions should decrease respectively by more than 15% over 2015; the concentration of $PM_{2.5}$ in cities at the prefecture level and above which fail to meet relevant standards should decrease by more than 18% over 2015, the proportion of good air quality days in cities at the prefecture level and above should reach 80%, and the proportion of days with heavy pollution and above should decline by more than 25% over 2015; provinces which have attained targets set forth in the 13th Five-Year Plan ahead of schedule should maintain and consolidate improvement results; those have not should ensure attainment of binding targets in full swing; ambient air quality improvement targets for Beijing should be further raised on the basis of those set forth in the 13th Five-Year Plan. Key areas include Beijing-Tianjin-

Hebei region and surrounding areas, the Yangtze River Delta, Fenhe and Weihe Plains, and so forth.

3.5 Ecosystem and Biodiversity Conservation

3.5.1 Positive progress has been made in ecological protection red lines drawing and conservation

The 2017 CCICED Policy Recommendations states that considerable attention has been given to environment-economy relationships in China. In the longer-term, social-environmental linkages may be among the most significant. This is the case for environmental related social unrest ("Not in My Back Yard"), management of nature protected areas, ecological redlining, climate adaptation, urban migration, and relationships of poverty reduction and the environment.

According to Several Opinions of General Office of the Communist Party of China and General Office of the State Council on Drawing and Strictly Observing Ecological Protection Red Lines released in February 2017, Beijing-Tianjin-Hebei region and provinces (municipalities directly under the Central Government) along the Yangtze River Economic Belt should draw ecological protection red lines by the end of 2017; other provinces (autonomous regions and municipalities directly under the Central Government) should draw ecological protection red lines by the end of 2018; and national ecological protection red lines should be drawn and an ecological protection red line system basically established by the end of 2020.

Plans of 15 provinces, including three provinces (municipalities) in Beijing-Tianjin-Hebei region, 11 provinces (municipalities) along the Yangtze River Economic Belt and Ningxia Hui Autonomous Region, for drawing ecological protection red lines have been approved by the State Council while those of other 16 provinces, including Shanxi, have been preliminarily developed. Based on scientific assessment, department coordination, planning coordination, regional coordination and balanced development of land and sea, departments such as MEE and NDRC, together with provinces concerned, have jointly completed delineation of ecological protection red lines for the above-mentioned 15 provinces upon expert demonstration and review by the inter-ministerial coordination leading group, through a combination of top-down and bottom-up methods and with national top-level design guiding local organizations.

A national ecological protection red line monitoring platform has been built and launched. In October 2017, NDRC officially approved the national ecological protection red line monitoring platform project, which, with a total investment of RMB 286 million, covers a total floorage of approximately 10,000 square meters and will be completed by the end of 2020. The design of a national ecological protection red line monitoring database has been completed, which has 67 data entries in four categories, with a total data size of 23.6TB. Some results have been verified and applied in "Green Shield 2017" Special Action for Supervision and Inspection of National Natural Reserves (hereinafter referred to as "Green Shield 2017"Special Action).

Supporting administrative policies have been developed for ecological protection red lines. Guidelines for Drawing Ecological Protection Red Lines and other guiding documents have been issued. Efforts have been made to study and draft measures for administration of ecological protection red lines and define requirements, principles and regulatory framework for administration of ecological protection red lines; so far, an exposure draft has been completed and will be issued to departments concerned and local governments for comment.

Additionally, local governments have been assigned to take main responsibility for drawing and observing ecological protection red lines, and should hold accountable those who break them, so as to ensure that these red lines are well drawn and work well. Delineation and implementation of ecological protection red lines will be incorporated into central environmental inspection.

3.5.2 "Green Shield 2018" provides strong guarantee for ecological protection in the Yangtze River basin

Seven departments, including MEE, MNR, MWR, Ministry of Agriculture and Rural Affairs, National Forestry and Grassland Administration, Chinese Academy of Sciences and SOA, had a video conference in Beijing on March 27 to deploy "Green Shield 2018" Special Action for Supervision and Inspection of Natural Reserves (hereinafter referred to as "Green Shield 2018" Special Action).

"Green Shield 2018" Special Action specifically includes: "reviewing" the rectification of problems identified in "Green Shield 2017" Special Action; resolutely investigating new violations in natural reserves and focusing on failures to well fulfill the responsibility for administration of national natural reserves; strictly supervising and handling identification and rectification of problems with natural reserves; etc.

A sound working mechanism has been established to ensure smooth development of the campaign. Inspection has been carried out against 469 national natural reserves and 847 provincial natural reserves to have a full knowledge of these reverses, with a ledger of problems built and prominent problems precisely solved, so as to promote higher level of administration in local natural reserves at all levels and facilitate healthy and sustainable development of natural reserves.

"Green Shield 2017" Special Action is a campaign featuring the widest scope of inspection, the most problems identified and handled, the most rigorous rectification and the strictest accountability system since China set up natural reserves. As at the end of 2017, a total of 20,800 clues to problems with 446 national natural reserves have been investigated and handled, with more than 1,100 people held accountable. Over 60% of the problems have been rectified, and a ledger has been built for the rest, which is being rectified.

Firstly, a long-term mechanism for ecological conservation of the Yangtze River should be established to resolutely contain sabotage to the ecological environment of the Yangtze River. Supervision and inspection should be strengthened and based on incorporation of national and provincial natural reserves into "Green Shield 2018" Special Action, city-level and county-level natural reserves at the main stream and tributaries of the Yangtze River and other natural reserves (tourist attractions, forest parks, wetland parks, etc.) at all levels should be included in the scope of supervision and inspection, so as to comprehensively and thoroughly identify violations sabotaging the ecological environment of natural reserves.

Secondly, importance should be attached to identifying problems with natural reserves at the main stream and tributaries of the Yangtze River, such as illegal encroachment on natural reserves, terminal construction, dredging and sand excavation, industrial development, mining, fishing of aquatic wildlife, encroachment on and damage to wetland, tourism development and hydropower development within core and buffer zones of natural reserves, a detailed ledger of problems should be built and the system of filing and cancellation upon rectification should be implemented to ensure thorough rectification.

Thirdly, violations regarding natural reserves should be handled seriously. Responsible units and individuals should be dealt with seriously and urged to rectify those violations timely and thoroughly. In case of ecological destruction resulting from any failure to well fulfill the responsibility for administration of natural reserves, local governments and departments concerned should be held accountable.

3.5.3 Ecological environment protection audit forces green development

On June 19, 2018, State Auditing Administration published findings from ecological audit of the Yangtze River Economic Belt. Fiscal funds of RMB 251,824 million were invested in ecological environment protection of the Yangtze River Economic Belt in 2016 and 2017. Progress in ecological environment protection is accompanied with some problems. With respect to fund use, eight local government authorities and affiliated units misused ecology-earmarked funds of RMB 25,804,900 from December 2013 to January 2018. With respect to resources exploitation, as at the end of 2017, construction of 930 small hydropower stations in eight provinces began without environmental assessment, and overexploitation dried up 333 rivers, stretching 1,017km in total. With respect to pollution control, 118 municipal wastewater treatment plants in sensitive areas in nine provinces failed to meet Grade 1A standard as at the end of 2017.

This audit is the first of its kind carried out by State Auditing Administration against the Yangtze River Economic Belt, for which a specialized audit report is released. Economically, the audit focuses more on ecological protection and restoration. In order to strengthen supervision through auditing and promote implementation of auditing decisions and rectification by audited units, State Auditing Administration will pay return visits after issuing auditing conclusions under relevant provisions. State Auditing Administration has issued an audit report according to law for problems identified through auditing and requested local governments concerned to publish rectification results after the rectification deadline.

In view of environmental protection audit practice in recent years, such audit is still in its infancy and is basically limited to audit of management and use of funds, without going deeper to further probe into root causes of environmental pollution, prevention and control measures, system establishment and fulfillment.

In the future, ecological environment audit should focus on the following aspects: Firstly, it should combine audit of ecology-earmarked fund use with pollution control results. Secondly, it should focus on construction and monitoring of key ecological environment projects. Thirdly, it should focus on fulfillment of ecological environment responsibilities by different departments. Fourthly, it should focus on direct and main accountability of leaders and cadres for development of "green economy". Fifthly, it should focus on the overall efficiency of ecology-earmarked funds.

3.6 Energy, Environment and Climate

3.6.1 Optimizing and adjusting the energy structure

As the Outline of the 13th Five-Year Plan states, China should deepen the energy revolution, energetically promote the reform of ways to produce and use energy, optimize energy supply structure, increase energy utilization efficiency and establish a clean, low-carbon, safe and efficient modern energy system.

2018 National Energy Work Conference put forward the new energy security strategy for the first time. The new strategy sets forth promoting the reform of energy consumption, supply, technology and system, strengthening international cooperation roundly, efficiently utilizing international resources and seeking to realize energy security in an open China. It focuses on accelerating the reform of energy development quality, efficiency and impetus and aims at maintaining high-quality energy development. Great progress has been made in energy work since the first half of 2018:

Firstly, the construction of a green and diversified energy supply system has been expedited. In the first half of 2018, non-fossil energy sources, such as hydropower, nuclear power, wind power and solar power, contributes to 25.2% of the total generating capacity, representing a 0.3% increase over the same period last year. Gas productivity construction has been steadily advanced at home, for example, the western Sichuan shale gas project in Chengdu, and the construction of key Coalbed Methane development projects has been quickened. Clean coal use is speeding up, the upgrading and demonstration of the coal-based oil and gas industry are steadily advancing and in particular, ultra low emission and energy conservation transformation of coal-fired units set forth in the 13th Five-Year Plan is expected to be completed ahead of schedule.

Secondly, the supply-side structural reform continues to show positive results. Efforts have been made to orderly develop high-quality coal production capacity in the principle of reduced capacity replacement, force faster exit of inefficient and low-quality production capacity and further increase the proportion of advanced coal production capacity. Total coal yield in Shanxi, Shaanxi and Inner Mongolia has increased to account for 70% of national output, and the number of coal mines nationwide has reduced from 12,000 at the end of the 12th Five-Year Plan to about 7,000. The sequence of coal-fired power planning and construction continues to be optimized, increase of new production capacity has been strictly controlled and outdated coal-fired power production capacity is being eliminated in a faster manner. Power consumption structure is being constantly optimized.

Thirdly, transformation of old impetus to new one has speeded up. From the perspective of power growth drive, power consumption growth rates in equipment manufacturing sectors in the secondary industry, such as medicine, general equipment, automobile, electric machinery and equipment, and computer/communication, have all exceeded or approached 10%. Power consumption by services such as information transmission/software technology in the tertiary industry has increased by 25.5% on a year-on-year basis, driving a year-on-year increase of 12.4% in total power consumption by emerging industries, which represents a 2.2% increase over the same period last year.

Fourthly, the cultivation of new energy consumption models and formats is accelerating. Power consumption growth in fields such as environmental protection and electric power replacement has become a new drive of power consumption growth in society at large.

3.6.2 Strengthening energy conservation and increasing energy efficiency

In order to enhance energy conservation by key energy consumption units and increase energy use efficiency, seven ministries and commissions, including NDRC, Ministry of Science and Technology (MOST), People's Bank of China (PBC), State-owned Assets Supervision and Administration Commission (SASAC), General Administration of Quality Supervision, Inspection and Quarantine (AQSIQ, now State Administration for Market Regulation), National Bureau of Statistics and China Securities Regulatory Commission (CSRC), have jointly revised the Administrative Measures for Energy Conservation by Key Energy-using Entities (hereinafter referred to as the Measures) and the revised Measures has come into force on May 1, 2018. The new Measures consists of 37 articles, 13 ones more than its predecessor, with the document size increasing by 1.5 folds, indicating a major revision.

Content of the revised Measures:

Firstly, a sound system of joint incentives for good faith and joint punishments for lack of good faith should be established, and it should be explicitly specified that departments in charge of energy conservation administration in people's governments at the provincial level and above, together with other departments concerned, should give incentives for good faith to key energy-consuming units and individuals that have made remarkable achievements in energy conservation and therefore been commended and rewarded.

Secondly, price policies conducive to energy conservation should be implemented and key energy-using entities should be encouraged to carry out power demand-side management, energy performance contracting (EPC) and voluntary commitment to energy conservation. As many as 11 enterprises in the cement industry, including Handan Jinyu Taihang Cement Co., Ltd., have been included in the List of 100 Energy-consuming Units Voluntarily Committed to Energy Conservation, which has been recently released by NDRC.

Thirdly, either differentiated tariff or step tariff should be applied to key energy-using entities in high energy-consuming industries based on their energy consumption indexes and other conditions. Power consumption by cement manufacturers for production has been subject to the policy of step tariff based on the integrated power consumption level of a comparable clinker (cement) since 2016.

Fourthly, departments in charge of energy conservation administration at all levels, together with other departments concerned, should provide key energy-consuming units included in the List of Energy Efficiency "Leaders" with priority support to improve energy efficiency through energy conservation technology transformation and information-based energy management, popularize advanced experience and drive an overall increase in the energy efficiency throughout the industry. In 2018, MOHURD has deepened the work in promoting the improvement of building energy efficiency and developed medium- and long-term development roadmaps for improvement of building energy efficiency by 2020, 2035 and the middle of this century.

Provinces and autonomous regions should promote energy conservation and energy efficiency improvement based on their respective conditions. For example, Beijing Municipal Commission of Development and Reform has released the Notice on Creating Energy Efficiency Leaders of 2018, which helps further raise the level of energy efficiency and energy conservation administration in Beijing, and organized the creation of Energy Efficiency Leaders of 2018.

Take Shanxi, a major province of energy, for another example. In 2018, energy consumption per unit of gross regional product (GRP) there has decreased by 3.2% over 2017, and the total energy consumed is less than 211.83 million tons of standard coal; energy consumption per unit of industrial added value of enterprises above a designated scale has decreased by 3.9% over 2017. A batch of key energy conservation projects should be implemented. Actions on evaluation and assessment of target responsibilities of the Top 100, Top 1000 and other key energy-consuming units should be implemented. An energy efficiency leader system should be established for high energy-consuming industries. In key energy-consuming industries such as steel, power, non-ferrous metals, coking and cement industries, an implementation plan for the energy efficiency leader system should be developed based on the reality of these industries to establish a benchmark of energy efficiency, guide key energy-consuming units to be aligned to advanced units and promote continuous improvement of industrial energy efficiency. An online energy consumption monitoring system should be built for key energy-consuming units and the construction of the system should be promoted.

3.6.3 Coordinating national efforts to cope with and adapt to climate change

China has been actively engaging in negotiations under the United Nations Framework Convention on Climate Change (UNFCCC), and has first proposed the scheme of intended nationally determined contributions (INDCs) to addressing climate change, guided and promoted the conclusion of key outcome documents such as the Paris Agreement and Fiji Momentum for Implementation. China has creatively established a multilateral ministerial negotiation platform, established the "BASIC" ministerial negotiation and coordination mechanism with India, Brazil and South Africa, set up the Like Minded Group of Developing Countries(LMDC) coordination mechanism with developing countries and initiated the mechanism of ministerial conference on climate actions with Canada and the EU. Meanwhile, China has been actively participating in negotiations outside the UNFCCC and has mobilized and brought into play synergy of multiple channels, which helps further build up the image of China as a responsible world power. Negotiations on rules for implementation of the Paris Agreement will be completed in 2018, in which we should continue to play an active and constructive role. We will also energetically promote the South-South cooperation and support other developing countries in coping with climate change.

China has set up a national leading group on climate change, energy conservation and emission reduction, which is led by Premier of the State Council and consists of leaders of more than 20 ministries and commissions. Different provinces, municipalities and autonomous regions have accordingly set up leading groups on climate change and established their own departmental labor division coordination mechanisms, thus forming a working system on climate change where the national leading group on climate change is responsible for overall planning and coordination, departments in charge of climate change are responsible for implementation, different departments cooperate with each other and the linkage between central and local authorities is guaranteed.

China has included carbon intensity reduction and other climate change targets in the 12th and 13th Five-Year Plans, established a responsibility system and strengthened decomposition, implementation and assessment of targets. China has organized capacity building and personnel training for better combating climate change and constantly raised the level of work of departments concerned at all levels on climate change. China has established a national panel on climate change and continuously enhanced the construction of national supporting teams against climate change, so that the research level and capacity of national and local supporting institutions continue to improve, providing strong support for policy making at all levels.

China will further strengthen work coordination and policy synergy against climate change, promote the construction of local working teams combating climate change, continue to coordinate and deepen the low-carbon pilot project and advance the work on adaptation to climate change. At present, we are studying and implementing the implementation plan for goals to be achieved by 2030, and exploring a low-carbon development strategy suitable for China by the middle of this century.

3.6.4 Promote the construction of a national carbon market

Establishing a national carbon emission permit trading market is a key task of the ecological civilization system reform and an important policy tool to control greenhouse gas emissions. The Plan for the Construction of a National Carbon Emission Permit Trading Market (for the Power Generation Industry) was released and enacted in December 2017, establishing a national carbon emission permit trading system first in the power generation industry. Statistics show that there are more than 1,700 enterprises in the power generation industry, emitting over 3 billion tons of carbon dioxide equivalence (at present, the world's largest EU carbon market emits approximately 2 billion tons of carbon dioxide equivalence).

Next, China should strengthen carbon market administration system construction, infrastructure construction and capacity building. It should promote the enactment of the Interim Regulations on Administration of National Carbon Emission Permit Trading and relevant supporting systems like allowance allocation. It should accelerate the construction of a data reporting system, registration system, trading system and settlement system. Also, it should organize capacity building activities for all kinds of market players.

3.7 Pollution Prevention and Control

3.7.1 Air pollution prevention and control

Since 2018, in order to ensure the victory in defending the blue sky, the State Council promulgated the Three-Year Action Plan on Defending the Blue Sky on June 27, injecting strength into further enhancing air pollution control. Meanwhile, MEE has released the implementation plan for air pollution prevention and control in key area such as Beijing-Tianjin-Hebei region and surrounding areas, the Yangtze River Delta and Fenhe and Weihe Plains, providing effective guidance for air pollution prevention and control in 2018.

From June 11 to July 8, 2018,MEE carried out two rounds of strengthened inspection of "2+26" cities in Beijing-Tianjin-Hebei region and surrounding areas. Despite remarkable achievements made by these cities in air pollution control, 5,204 problems with atmospheric environment were identified. The most problems were found in Quyang County, Hebei, followed by Tongzhou District, Beijing; a rebound in the number of problems was noted in the second round of inspection of Zhaoxian County, Hebei and the urban area of Jincheng, Shanxi, while Huixian, Henan was identified to failure to rigorously carry out pollution control. 119 problems with atmospheric environment were identified in Quyang County, Hebei. 34 small coal-fired boilers in 28 enterprises (units) were not eliminated as required; some enterprises ran no pollution control facilities or had prominent unorganized emissions.

2018—2019 strengthened inspection of key areas of the Blue Sky Protection Campaign continued on August 7 and 200 inspection teams examined 208 counties (prefectures and districts) in Beijing-Tianjin-Hebei region and surrounding areas as required and found 103 air pollution-related environmental problems. See Box 3-1 for specific information on inspection.

Box 3-1 Strengthened Inspection of Key Areas of the Blue Sky Protection Campaign 2018—2019

1. Inspection teams inspected 1,320 air pollution-related "poorly managed and polluting small" enterprises in the list and found inadequate rectification in two of them, accounting for 0.2%; they also found an unlisted "poorly managed and polluting small" enterprise.

Inspection teams found inappropriate rectification in two air pollution-related "poorly managed and polluting small" enterprises in the list, one located in Zhengding County, Shijiazhuang, Hebei and the other in Pingyuan County, Dezhou, Shandong.

Inspection teams found that Wang Fang Xin Furniture Factory (formerly Hebei Huazhong General Furniture Co., Ltd.), located in Nangong, Xingtai, Hebei, was an unlisted air pollution-related "poorly managed and polluting small" enterprise.

2. 403 coal-fired boiler enterprises in the list should be eliminated, but a coal-fired boiler was founded not torn down, accounting for 0.2%; five more unlisted boilers that should be eliminated were found.

Inspection teams found that a listed boiler (less than 1 steam ton) at Huayuan Hotel in Dacheng County, Langfang, Hebei, which should be eliminated, was not torn down.

Inspection teams found five more unlisted boilers that should be eliminated, including one (0.2 steam ton) at Yujingzhou Car Dealer in Zhuozhou, Baoding, Hebei, one (less than 0.1 steam ton) at Changsheng Hardware Co., Ltd. in Yunhe District, Cangzhou, one (0.15 steam ton) at Qinghe No.3 Gas Station of Xingtai Qinghe Branch of Sinopec, one (0.1 steam ton) at Tongda Hotel in Qinghe County and one (0.1 steam ton) at Dongdong Restaurant in Ningjin County.

3. 13 industrial enterprises were found have no air pollution prevention and control facilities

Among these enterprises, two were found in Beijing, one in Shunyi District and the other in Mentougou District; five were found in Hebei, one in Luquan District and one in Yuanshi County, Shijiazhuang, one in Langfang Development Zone and one in Xianghe County, Langfang, and one in Xinhua District, Cangzhou; two were found in Shanxi, one in Gujiao, Taiyuan and the other in the urban area of Yangquan; and four were found in Henan, one in Fuxiang District, Kaifeng, one in Anyang County, Anyang City, one in Wuzhi County, Jiaozuo and one in Mengzhou.

4. 13 industrial enterprises were found fail to run air pollution prevention and control facilities normally

Among these enterprises, one was found in Shunyi District, Beijing; two were found in Dongli District, Tianjin; four were found in Hebei, one in Langfang Development Zone, Langfang, one in Boye County, Baoding, one in Yunhe District, Cangzhou and one in Jizhou Disctrict, Hengshui; one was found in Gaoping, Jincheng, Shanxi; and five were found in Henan, one in Jinshui District, Zhengzhou, one in Xinmi, one in Huang County, Anyang and two in Urban-Rural Integration Demonstration Zone, Hebi City.

5. Inadequate VOCs control was found in 11 enterprises

Among these enterprises, one was found in Pinggu District, Beijing; seven were found in Hebei, one in Wuji County, Shijiazhuang, one in Lutai Economic Development Zone, Tangshan, one in Anci District, Langfang, one in Nanpi County, Cangzhou, one in Botou City and two in Pingxiang County, Xingtai; and three were found in Henan, two in Jinshui District, Zhengzhou and one in Shunhe Hui District, Kaifeng.

6. Unorganized industrial dust emission was found in 15 enterprises

Among these enterprises, one was found in Tongzhou District, Beijing; 10 were found in Hebei, one in Luquan District, one in Luancheng District, one in Wuji County, one in Yuanshi County and one in Xingtang County, Shijiazhuang, one in Guangyang District, Langfang, one in Yunhe District, one in Qing County, one in Dongguang County and one in Botou Prefecture, Cangzhou; one was found in Changzhi County, Changzhi City, Shanxi; and three were found in Henan, one in Huang County, Anyang, one in Muye District, Xinxiang and one in Changyuan County.

7. An enterprise was found fail to take dust reduction measures for open-pit mines

Fubao Mining Co., Ltd. in Laishui County, Baoding, Hebei had no dust reduction measures for open-pit mines.

8. 26 enterprises were found with poor dust management at construction sites

Among these enterprises, three were found in Beijing, one in Shunyi District, one in Chaoyang District and one in Mentougou District; two were found in Tianjin, one in Jinghai District and the other in Wuqing District; 11 were found in Hebei, one in Luancheng District, one in Jinzhou and one in Xinji, Shijiazhuang, two in Yunhe District and one in Xinhua District, Cangzhou, two in Qiaoxi District, one in Ningjin County, one in Nangong and one in Wei County, Xingtai; three were found in Shanxi, one in Gujiao, Taiyuan, one in Changzhi County and one in Qinyuan County, Changzhi City; one was found in Dongchangfu District, Liaocheng, Shandong; and six were found in Henan, one in Jinshui District and one in Guangcheng Hui District, Zhengzhou, one in Linzhou, Anyang, two in Huojia County, Xinxiang and one in Taiqian County, Puyang.

9. 15 enterprises were found fail to take dust control measures for material yards

Among these enterprises, one was found in Tongzhou District, Beijing; six were found in Hebei, one in Xingtang County, Shijiazhuang, one in Sanhe, Langfang, one in Laishui County, Baoding, one in Gaobeidian City, one in Ningjin County, Xingtai and one in Handan Economic and Technological Development Zone; three were found in the suburb of Yangquan, Shanxi; and five were found in Henan, one in Zhongmu County, Zhengzhou, two in Hebi Economic and Technological Development Zone, one in Yuanyang County, Xinxiang and one in Puyang County, Puyang City.

As time goes on, the autumn and winter approach, so MEE has developed the Action Plan for Comprehensive Control of Air Pollution in Beijing-Tianjin-Hebei Region and Surrounding Areas in the Autumn and Winter of 2018—2019 and solicited comments from all parties. Then, MEE will, together with local authorities and departments concerned, steadily promote clean energy heating in northern China, accelerate elimination of small coal-fired boilers, constantly facilitate bulk coal pollution control, focus on upgrading and reform of key industries, expedite motor vehicle pollution control and strengthen joint pollution prevention and control in key areas.

Additionally, advanced technologies have effectively supported air pollution control. By launching a "clairvoyance project" in 2018, MEE has divided all administrative regions in the "2+26" cities in Beijing-Tianjin-Hebei region and surrounding areas into grid units measuring 3km×3km, and screened out through satellite remote sensing 3,600 grid units with relatively high annual average concentration of $PM_{2.5}$ as hot spot areas to watch, so as to timely identify and precisely solve air problems. With satellite remote sensing, this project can calculate the average $PM_{2.5}$ intensity in each grid unit in the previous year based on meteorological data and air quality monitoring data, and then work out the level of emissions under stagnant weather conditions without regard to the impact of factors such as meteorological transmission, precipitation and topographic countercheck, thus completing the screening. For grid hot spots warned for three consecutive times within a year or for six times in total, MEE will urge local authorities to improve the environment by means of notification, sending working groups and publicly interviewing heads of local governments.

3.7.2 Water pollution prevention and control

In April 2018, the first meeting of the Central Financial and Economic Affairs Commission was held to deliberate ideas and measures for a victory in the three tough battles. According to the meeting, to win the tough battle of pollution prevention and control, we should make clear the objective and task, i.e., to remarkably reduce total key pollutant emissions and improve the overall quality of ecological environment by 2020. We should fight in several symbolic key battles, secure the victory in the Blue Sky Protection Campaign, win tough battles such as control of pollution caused by diesel trucks, control of urban black and odorous water bodies, comprehensive control of the Bohai Sea, conservation and restoration of the Yangtze River, protection of water sources and control of agricultural and rural pollution, and ensure that obvious effects will be achieved in three years. We should make major measures to win the tough battle of pollution prevention and control as detailed as possible, respect rules and insist on the bottom line thinking. Party committees at all levels and Party leadership groups (Party committees) of different departments should give priority to pollution prevention and control, urge lower levels to implement the work and mobilize all segment of society to engage in pollution prevention and control. We should stick to prevention and control from the source, adjust the "four structures" and ensure "four decreases and four increases". Firstly, we should adjust the industrial structure by decreasing excess and outdated production capacity and increasing new drives of growth. Secondly, we should adjust the energy structure by decreasing coal consumption and increasing the use of clean energy. Thirdly, we should adjust the transportation structure by decreasing highway transportation and increasing railway transportation. Fourthly, we should adjust the agricultural input structure by decreasing the use of chemical fertilizers and pesticides and increasing the use of organic fertilizers. We should insist on overall consideration and systematic planning, reflect differentiation and the system of reward and punishment, and avoid affecting people's life.

Among the five "water battles", we should start with the control of black and odorous water bodies. MEE, together with MOHURD, has launched a special campaign on control of black and odorous water bodies, which covers 36 key cities and some other prefecture-level cities nationwide, and will carry out a special inspection of the first batch of eight provinces, autonomous regions and municipalities, including Guangdong, Guangxi and Hainan, which will start in September and October, 2018. Those failing to effectively rectify problems identified will be held accountable by central environmental inspectors.

In recent years, comprehensive control of ecological environment around the Bohai Sea has aroused public concern. Great pressure has been posed on the ecological environment around the Bohai Sea due to its poor self-purification capability and large-scale land reclamation from sea by the government, overexploitation of marine resources and large-scale mariculture over the past years. Following this round of organizational reform, the duty of marine environment governance has been transferred from SOA to MEE. MEE will insist on balanced development of land and sea, advance pollution control and ecological protection at the same time, and work with departments, provinces and cities concerned to develop an action plan on comprehensive control of the Bohai Sea as soon as possible. In other words, MEE will intensify efforts to inspect and rectify sewage outlets to the sea, clean up illegal and unreasonable outlets, and focus on the control of ship and port pollution as well as the construction of refuse transfer and disposal facilities.

Besides, water bodies as sources of drinking water will top the list of the "four water bodies" subject to water pollution prevention and control. MEE and MWR have jointly launched a special campaign on environmental protection of centralized drinking water sources nationwide. They have set up the following milestones: to complete cleanup and rectification of county-level surface water sources along the Yangtze River Economic Belt and those at the prefecture level and above in other areas by the end of 2018; and to complete cleanup and rectification of county-level surface water sources in all other areas by the end of 2019. According to statistics, as many as 2,466 surface water sources are covered by this campaign, including 1,161 county-level water sources along the Yangtze River Economic Belt. From the perspective of the type of problems, illegal discharge of sewage by industrial enterprises is prominent.

3.7.3 Soil pollution prevention and control

With respect to soil pollution prevention and control, MEE will steadily advance the "Clean Soil Campaign". Specific measures mainly include: comprehensively implement the Soil Pollution Prevention and Control Action Plan (hereinafter refferred to as the Action Plan), strengthen soil pollution risk control; deeply advance detailed surveys on soil pollution and safeguard the safety of agricultural land and construction land; accelerate the construction of pilot zones for comprehensive prevention and control of soil pollution and pilot application of soil pollution control and remediation technology; enhance prevention and control of solid waste pollution and expedite classified disposal of refuse.

In 2018, MEE developed the Provisions on Evaluation and Examination of the Implementation of the Soil Pollution Prevention and Control Action Plan, guiding annual assessment and mid-term examination of the implementation of the Action Plan by provincial, autonomous regional and municipal people's governments during 2018—2020. From 2019 to 2021, MEE will evaluate the implementation of the Action Plan by local governments over the previous year and the completion of priorities in soil pollution prevention and control at the beginning of each year.

The evaluation and examination cover the completion of soil pollution prevention and control objectives and the completion of priorities in soil pollution prevention and control. Specifically, annual evaluation covers the completion of priorities in soil pollution prevention and control; while the mid-term examination covers the completion of soil pollution prevention and control objectives and the completion of priorities in soil pollution prevention and control.

Priorities in soil pollution prevention and control include: detailed surveys on soil pollution, prevention from the source, classified management of agricultural land, construction land access management, pilot demonstration, fulfillment of responsibilities of all parties and public engagement. Soil pollution prevention and control objectives include safe utilization ratio of polluted farmland and safe utilization ratio of polluted plots.

In order to guide local governments to entrust third parties to evaluate pollution control and remediation results, implement the Action Plan, complete soil pollution control and remediation objectives and tasks, and safeguard the quality of agricultural products and the safety of the living environment, MEP printed and released the Guidelines for Technical Evaluation of Soil Pollution Control and Remediation Results (for Trial Implementation) in December 2017. The Guidelines plays an important guiding role in regulating the evaluation of soil pollution control results by third parties entrusted by provincial (autonomous regional and municipal) people's governments or departments concerned. Meanwhile, it lays a foundation for China to evaluation the results of actions against soil pollution taken by different provinces, autonomous regions and municipalities.

3.8 Governance and Rule of Law

3.8.1 Major organizational reform in ecology and environmental protection

Article 2 of the 2017 CCICED Policy Recommendations proposes to build Ecological Civilization co-management systems to improve ecological and other services in nature reserves, parks, ecological redline areas, and in other public lands that sometimes are considered to be of low value. As CCICED's second policy recommendation in 2017 states, "a collaborative management system for ecological civilization should be established".

A large-scale reform of national and local government organizations was carried out in March 2018, with even radical adjustments in some fields, which is unprecedented in intensity, depth and breadth. Functions of MEP are further upgraded and expanded, which fully reflect an "integration of pollution governance and ecological regulation". The newly established MEE will act as a component department of the State Council that undertakes MEP's all duties, NDRC's duties of tackling climate change and reducing emissions, MLR's duties of supervising and preventing groundwater pollution, MWR's duties of formulating water functional zoning, managing sewage outlet setting and protecting watershed environment, MOA's duties of supervising and guiding agricultural non-point source pollution control, SOA's duty of marine conservation and the duty of Office of South-to-North Water Diversion Project Commission, the State Council of environmental protection in the South-to-North Water Diversion Project area.

It should be noted that the incorporation of NDRC Department of Climate Change into MEE is actually a specific move to synergize work such as pollutant emission reduction and greenhouse gas emission reduction, combating climate change, etc. Combating climate change involves energy utilization and carbon dioxide emission, which go beyond pollution control as carbon dioxide itself is not a pollutant, but a matter causing changes to greenhouse gases. Now carbon dioxide emission is being regulated by MEE, making the regulation over emissions more comprehensive and making it easier to achieve the overall effect of ecological civilization construction.

At the level of ministerial comprehensive coordination for energy conservation and emission reduction, the State Council decided to adjust units and personnel composing the National Leading Group to Address Climate Change and Energy Conservation & Pollutant Discharge Reduction based on its organizational structuring, staff turnover and work needs. The head of the Leading Group is Li Keqiang, Premier of the State Council, and deputy heads are Han Zheng, Vice Premier of the State Council, and Wang Yi, State Councilor. Following such adjustment, members include Ding Xuedong, Deputy Secretary-General of the State Council and leaders of departments such as Ministry of Foreign Affairs, NDRC, Ministry of Education, MOST, Ministry of Industry and Information Technology, Ministry of Civil Affairs, Ministry of Justice, Ministry of Finance, MNR, MEE, MOHURD, Ministry of Transport, MWR, Ministry of Agriculture and Rural Affairs, Ministry of Commerce, Ministry of Culture and Tourism, National Health Commission, PBC, SASAC, State Administration Taxation, State Administration of Market Regulation, National Bureau of Statistics, International Development Cooperation Agency, National Government Offices Administration, Chinese Academy of Sciences, China Meteorological Administration, NEA, National Forestry and Grassland Administration, National Railway Administration and Civil Aviation Administration of China. The specific work of the Leading Group is undertaken by MEE and NDRC according to their respective duties.

3.8.2 Emission permits

Implementing the permit system for controlling pollutants emission is an important part of the ecological civilization system reform and a central task of reforming and improving the stationary pollution source management system. Since the State Council printed and released the Implementation Plan for the Permit System for Controlling Pollutants Emission in 2016, MEP issued the Interim Measures for Pollutants Emission Permit in early 2018, and will promulgate the Catalog of Classified Management of Pollutant Emission Permits, print and release the general principles of the Technical Specifications for Application for and Issuance of Pollutant Emission Permits in Thermal Power and Paper-making Industries and Technical Guidelines for Self-Monitoring by Pollutant Emission Units, thus preliminarily building a relatively complete system of regulations and technical specifications to fully support the application for and issuance of pollutant emission permits.

A national pollutant emission permit management information platform has been basically built and put into operation, which undertakes application for and issuance of pollutant emission permits for enterprises in thermal power and paper-making industries, basically realizing a convenient, precise and efficient information-based management model.

At present, the work on application for and issuance of pollutant emission permits for thermal power and paper-making industries has been fully launched, for which local environmental authorities are actively deploying and organizing the implementation. As of May 18, 6,073 enterprises have registered on the national pollutant emission permit management information platform, pollutant emission permits applied for by more than 1,500 are awaiting approval and environmental authorities have issued 68 permits subject to national unified coding.

Generally, the issuance of pollutant emission permits is advancing orderly, but problems such as failure to pay enough attention ideologically or transfer the pressure in time and inadequate technical force still exist in some areas. Meanwhile, some local governments are perplexed by the heavy task of systems linking, which partly affects the work progress.

3.8.3 Thoroughly implementing environmental economic policies

Xi Jinping's report at the 19th National Congress of the Communist Party of China suggests that we should create a market-based system for green technology innovation, develop green finance, and spur the development of energy-saving and environmental protection industries as well as clean production and clean energy industries. Green finance will be an important aspect of financial reform in China.

CCICED has played a leading role in the introduction, propelling and demonstration of green finance. In 2014, it set up a Green Finance Task Force. Thus the year of 2014 saw the beginning of green finance in China. On August 30, 2016, the Central Leading Group for Comprehensively Deepening Reforms convened a meeting, and deliberated and adopted the Guiding Opinions on Building a Green Financial System. The Opinions gave a comprehensive plan for green finance reform and firstly proposed to include green finance into G20 Agenda. It was at 2016 G20 Hangzhou Summit, green finance was on the agenda as one of key topics, setting of an upsurge of green finance. The Opinions clearly specifies that we should gradually establish and improve mandatory environmental information disclosure systems of listed companies and bond issuers.

Since 2016, green finance has shown a good trend of full-scale acceleration: on the one hand, top-level design and fundamental institutional arrangements are increasingly improved. The Guiding Opinions of the People's Bank of China and Other Six Ministries and Commissions on Building a Green Financial System steadily advances various basic work, for example, making uniform standard for defining green projects and for green finance. National green finance reform and innovation pilot zones are established in Zhejiang, Guangdong and other five provinces and regions to explore replicable and generalizable experiences. On the other hand, more and more green investment and financing instruments further optimizes the financial eco-environment, with fruitful results gained in international cooperation.

3.8.3.1 Green liability insurance

On May 7, 2018, MEE adopted in principle the Administrative Measures for the Compulsory Liability Insurance for Environmental Pollution (hereinafter referred to as the Measures) which proposed to establish compulsory liability insurance system for environmental pollution in sectors that are highly exposed to environmental risks. It is a strong measure and a concrete action to carry out the spirit of the 19th National Congress of the Communist Party of China, and it is necessary and important for establishing a sound green finance system.

By summing up pilot experiences, the Measures further standardizes compulsory liability insurance system for environmental pollution, and enriches market-based means for eco-environment protection, which is of positive significance to fight a tough battle against environmental pollution and make up for the deficiency of eco-environment in building a moderately well-off society in an all-round way.

As indicated by relevant data, in 2014, about 5,000 enterprises insured against environmental pollution liability; in 2015, 14,000 policies of insurance against environmental pollution liability were reported, with the premium of RMB 280 million and offering risk securities of RMB 24.421 billion; in 2016, the premium income from the insurance against environmental pollution liability was close to RMB 300 million, offering risk securities more than RMB 26 billion; in 2017, the insurance against environmental pollution liability offered risk securities of RMB 30.6 billion for more than 16,000 enterprises. The Measures will, on the basis of the pattern of "Further Environmental Protection", well develop and practice the system of compulsory environmental pollution liability insurance, introduce market-oriented professional forces and "evaluate and price" environmental risk, to realize internalization of external cost and improve effects of environmental risk regulation and compensation for damages. MEE will work together with relevant departments to properly carry out mechanism building, and enhance works to map out corresponding codes on intensifying "pre-event" prevention, "in-event" control and "post-event" handling, so as to further improve the pertinence and operability and ensure effective implementation of relevant measures.

3.8.3.2 Green bond

The green bond policies introduced in the first half of 2018 are as follows: PBC and CSRC jointly issued Guideline on Green Bond Assessment and Certification to standardize green bond assessment and certificate; Shanghai Stock Exchange proposes new policy requirements on corporation green bonds; and two preferential policies on green bond were put into implementation.

As substantive preferential policies such as financial discount, tax reduction and exemption have not yet been implemented in 2018, consumers' enthusiasm for green bond market goes down. Thus it is urgent to put substantive preferential policies into implementation. But from the perspective of regulatory dynamics, regulators are actively pushing on the implementation of preferential policies.

A total of 49 green bonds (excluding additional bonds and Green Asset-backed Securities) are issued in the first half of 2018, with the issuing scale up to RMB 590.56 billion, an increase of 48.48% and 812.91% y-o-y respectively. It is mainly attributed to the large increase in the issuing scale of financial bonds. While the issuance of corporation green bonds is scaled down on a y-o-y basis, showing that the green bond market is less attractive to consumers. Green bonds for energy conservation projects are most widely issued and money is raised to invest in industrial energy conservation and sustainable construction. Iron and steel enterprises and coal enterprises have also issued green bonds.In addition to financial bonds, Guangdong reports the largest issuance scale of green bonds. By June 30, 2018, 12 green bonds have been issued overseas by domestic entities, mainly by banks. Other entities have AAA credit qualifications, most of which are companies engaging in green industry.

3.8.3.3 About green credit

Green credit, referring to loans that are invested in green projects and environmental improvement, boasts a long loan period, low interest rates as well as reduction and exemption of fees, and focuses on providing medium and long-term financing services for sustainable infrastructure projects. China has gradually issued relevant policies to support the healthy development of green credit, and taken actions, such as expanding the scale of funds with fiscal discounts and establishing professional guarantee institutions, to boost the rapid development of green loans. Policy banks, including China Development Bank, New Development Bank of BRICS Countries and Asian Infrastructure Investment Bank, are among the most important providers of green credit, who mainly aim at programs in green fields, such as environmental governance and clean energy. In China, the fiscal discount rate of green loans is based on the benchmark lending rate of PBC in most cases, and the actual interest rate in some cases, but the upper limit is 3%. This means green credit has obvious advantages financing costs, and will become a financing mode favored by relevant enterprises.

China has witnessed the rapid growth of the size of the green financial market in recent years. Statistics show that the overall balance of various types of green financing had reached nearly RMB 9 trillion by the end of 2017, of which green credit accounted for over 95%, making it a major driver for the construction of ecological civilization and transformation of green development. China's total scale of green credit has grown steadily in recent years, according to the China Banking Regulatory Commission (CBRC). The green credit balance of 21 Chinese banks rose from RMB 5.20 trillion in 2013 to RMB 8.22 trillion in June 2017, with an annual growth rate of 13.98%. Notably, the defect rate of green credit is only 0.37%, far lower than the average defect rate of 1.69% of other loans during the same period, showing that the asset quality of green credit is high.

3.8.3.4 About green price

NDRC issued Opinions on Innovating and Optimizing the Price Mechanism on the Promotion of Green Development (hereinafter referred to as Opinions) in June 2018. The Opinions pointed out that market-oriented means should be fully used to advance the marketization of ecological and environmental protection, continuously optimize the price mechanism on resources and the environment, bring price leverage into better play in guiding the optimal allocation of resources, achieving ecological and environmental cost internalization, promoting society-wide conservation, and accelerating the development of green environmental protection industry, thus mobilizing the whole society to jointly facilitate green development and ecological civilization building. The Opinions put forward that: by 2020, the price mechanism and the price policy system beneficial to green development will be basically formed, so as to significantly boost resource conservation and ecological and environmental cost internalization. By 2025, the price mechanism satisfying the requirements of green development will be further improved.

The Opinions proposed 15 specific guiding opinions in four aspects, such as optimizing the water treatment charging policy, improving the charging mechanism on solid waste treatment, establishing the water-saving price mechanism, and optimizing the electricity price mechanism on the promotion of energy conservation and environmental protection. These guiding opinions include implementing the differentiated charging mechanism for enterprises, and encouraging local governments to formulate classified and differentiated charging standards according to the type, density and environmental credit rating of major pollutants in sewage discharged by enterprises, so as to help enterprises promote sewage pre-treatment and pollutant emission reduction. Other guiding opinions also include optimizing the differentiated electricity price policy. China should comprehensively clear up and cancel preferential electricity prices for high energy-consuming industries as well as other various unreasonable favorable price policies. The government should also strictly implement the differentiated electricity price policy in seven sectors, including iron alloy, carbite, caustic soda, cement, steel, yellow phosphorus and zinc smelting, and formulate higher prices for electricity consumption (including electricity capacity through market-oriented transactions) of obsolete and restricted enterprises. Local governments should promptly evaluate the implementation effect of differentiated and tiered electricity price policies, expand the scope of sectors in implementing differentiated and tiered policies according to the actual needs, raise the standard for price makeup, and help related industries strengthen technical transformation, improve the energy efficiency level, and step up eliminating backward production capacity.

3.8.3.5 About green finance project demonstration

To better push forward the green finance work, CCICED established a green finance demonstration project, i.e. the “environmental information disclosure system for listed companies”. It aims to combine the actual situation of China’s capital market through the research and summarization of international experience in implementing environmental information disclosure systems, and put forward policies and suggestions to improve the environmental information disclosure system for listed companies in China.

The environmental information disclosure work should be an obligation of all listed companies. China plans to establish a mandatory environmental information disclosure system for listed companies step by step, according to the plan for division of work given in the Guiding Opinions on Establishing the Green Finance System, which was jointly issued by seven ministries and commissions, including PBC, in 2016. The optimization of this system is facing many challenges, such as low willingness to environmental information disclosure from some major pollutant discharging units and disclosure without obeying regulations. Besides, stock exchanges, as a key execution unit of the system, while urging listed companies to disclose environmental information, fail to establish explicit index systems and achieve the efficiency and significance of information disclosure. The advancement of this demonstration project will resolve some short boards or difficulties that restrict environmental information disclosure of listed companies through such measures as establishing an appropriate reward and punishment mechanism, linking information disclosure to corporate refinancing, etc. The project demonstration result will provide an institutional basis for the comprehensive implementation of environmental information disclosure among listed companies for the next step. Since April, several listed companies, such as Gansu Shangfeng Cement, Shanxi Sanwei Group and Yancheng Huifeng Joint-stock, have been punished over serious environmental violation issues, attracting much attention to environmental violation issues of listed companies from the CSRC and MEE of China. On June 22, the CSRC issued an announcement that it will seriously rectify environmental information disclosure violations of listed companies.

According to the arrangements, step 1, CSRC requires listed companies to revise the content and format criteria of regular reports and implement voluntary disclosure by the end of 2017; step 2, it raises semi-mandatory disclosure requirements to all listed companies: key pollutant discharging units must disclose environmental information, and units without disclosing information must make explanations (before March 2018); step 3, all listed companies must implement environmental information disclosure (before December 2020).

3.8.4 Judicial reform drives ecological civilization governance

For environmental departments, forming joint forces through organizational function integration are undoubtedly important, but they expect more improvements in the environmental law enforcement in the whole society and an all-around way. These improvements include following up environmental legislation, optimizing the punishment mechanism, strengthening environmental judicial enforcement, improving environmental monitoring capabilities, changing the performance assessment of governmental officials, etc.

In 2018, the judicial authority issued the latest version of the General Provisions of the Civil Law. According to Article 9 of Basic Provisions of the law, the parties to civil legal relations shall conduct civil activities contributing to the conservation of resources and protection of environment. This is the first time that the General Provisions of the Civil Law has stipulated that civil activities should be consistent with resource conservation and environmental protection, which has become the green principle. This exactly coincides with the development of green lifestyle proposed at the 19th National Congress of the Communist Party of China .

On June 4, 2018, the Supreme People's Court of China (SPC) issued the Opinions on Deeply Studying and Implementing the Xi Jinping's Thought on Ecological Civilization and Providing Judicial Services and Guarantee for Ecological and Environmental Protection in a New Era (hereinafter referred to as Opinions), urging local courts to bring the judicial function of environmental resources into better play, and strengthen judicial services and guarantee through ecological civilization construction. The Opinions pointed out that more attention should be paid to develop innovative trial implementation modes, and promote the overall protection, system recovery, regional overall planning and comprehensive management of the ecology and environment. The Opinions stressed that more efforts should be made to solve outstanding ecological and environmental problems, and serve and guarantee pollution prevention and control and eco-security protection. To help win the tough battle against pollution prevention and control, SPC should judge cases relating to air, water, soil, solid waste, garbage disposal, noise and vibration pollution according to law, and strengthen the hearing of air pollution dispute cases in key regions, including Beijing-Tianjin-Hebei and neighboring areas, the Yangtze River Delta, the Fen-Wei Plain, etc., so as to provide strong judicial support for the blue sky defending war. Besides, more efforts should be made to hear water pollution dispute cases in key waters, such as Yangtze River, Yellow River, Poyang Lake, Dongting Lake and Taihu Lake, and promote treatment of black and odorous water bodies, thus practically safeguarding people's rights to live in a quiet and beautiful ecological environment where there are the blue sky, clear water and clean soil. SPC should protect marine natural resources and the environment and judge cases relating to the pollution and damage of the marine ecology according to law. SPC should make all-out efforts to serve the construction of beautiful countryside, implement the requirements of the rural rejuvenation strategy, and hear various types of cases relating to rural natural resource exploitation and utilization, the comprehensive management of rural living environment, agricultural ecological products and services, in an effort to promote the sustainable development of agriculture and rural affairs and the building of ecologically livable beautiful countryside. SPC should constantly improve its biodiversity protection capability, judge cases about the damage of biodiversity according to law, maintain the diversity of species and the ecosystem, and help improve the legal system of the biodiversity protection. SPC should rigorously guarantee the layout of the ecological security strategy, strictly stick to the red line of ecological protection, the bottom line of environmental quality and upper limits of resource utilization, hear cases involving key ecological function areas and ecological environment sensitive and vulnerable areas according to law, focus on ecological protection and recovery, and build an barrier for ecological security.

3.9 Regional and International Engagement

3.9.1 Greening Belt and Road initiative

Article 5 of the 2017 CCICED Policy Recommendations states as follows: China's green development approach, the UN 2030 SDGs, the Paris Agreement,

Biodiversity and Desertification global accords, and ecological civilization should become central features in the Belt and Road Initiative...Also, a Belt and Road Green Governance Mechanism including information disclosure, public participation and access to arbitration, should be established.

In order to advance the progress of greening the BRI, China, as the initiator of the initiative, took the lead in carrying out much fruitful work, such as the BRI eco-environment big data service platform for information sharing, the establishment of the BRI international coalition for green development together with the UNEP, the issuance of the Guiding Opinions on Promoting a Green Belt and Road Initiative and the Belt and Road Initiative Eco-environment Protection Cooperation Plan, the holding of themed exchange activities such as the International High-Level Dialogue on Ecological and Environmental Protection under the Framework of the Belt and Road Initiative, the founding of the Belt and Road Environmental Technology Exchange and Transfer Center (Shenzhen) and China-ASEAN Environmental Technology and Industry Cooperation and Exchange Demonstration Base (Yixing), and the promotion of the launch of the initiative on Performing Corporate Environmental Responsibility and Building a Green Belt and Road Initiative by related Chinese enterprises. At the same time, as Chinese companies are increasingly "going global", China's "green solutions" are also "carried around". China's several import and export chambers of commerce have established social responsibility indicators for foreign investment to guide enterprises to follow the principle of sustainable development. The China Textile and Apparel Association is a good example. At the same time, China has also promoted the environmental standards on the development of cobalt mine and rare earth to ensure the sustainability of mining activities at home and abroad.

Integrating the concept of green development into the BRI, and promoting BRI projects to bring about sustainable development effects both economically and environmentally have provided new impetus for the implementation of the 2030 Sustainable Development Goals. At present, 15 United Nations agencies and relevant Chinese institutions have signed agreements related to the BRI. The core of these agreements is to create synergies between the 2030 Agenda for Sustainable Development and the green Belt and Road Initiative.

As the focus of the BRI, the connectivity of infrastructure is a priority area for greening Belt and Road Initiative. After the BRI was released, China has strengthened cooperation with countries and regions along the route in highway, railway, aviation, shipping, energy, information infrastructure and other areas of infrastructure.

3.9.2 Making new progress in tackling climate change with active efforts

Article 5 of the 2017 CCICED Policy Recommendations states as follows: Strengthen global and regional green governance. It proposes that "China needs to start early to develop its own Mid-Century Climate Strategy and link it with others".

Climate change is a major global challenge facing humanity in the 21st century. Actively coping with climate change is not only an inherent requirement of China's own sustainable development and an important part of building a socialist modern strong power, it is also China's responsibility as it participates in and leads global climate governance and promoting the building of a community of shared destiny for mankind.

China has integrated climate change into its medium and long-term national economic and social development plan. It puts mitigation and adaptation to climate change at equal positions and pushes forward through all possible means such as law, administration, technology, and market. At present, China's renewable energy installed capacity accounts for 24% of the global total, and newly installed capacity accounts for 42% of the global increase. It is the world's largest country in energy saving and new and renewable energy utilization.

On December 19, 2017, NDRC organized a video teleconference on the launch of a national carbon emissions trading system, mobilizing and deploying the comprehensive implementation of the tasks and requirements in the Scheme and the construction of a national carbon emissions trading market.

The Scheme clearly states that the power generation industry (including cogeneration) is the first industry to launch a national carbon emissions trading system. The main participants are enterprises with annual emissions of 26,000 tons of carbon dioxide equivalent or more in the power generation industry or other economic organizations including other industry-owned power plants. The first batch of more than 1,700 companies involved in carbon trading, the total emissions exceeded 3 billion tons of carbon dioxide equivalent.

The Scheme pointed out that the ultimate goal of carbon market construction is to establish a carbon market with clear ownership, strict protection, smooth circulation, effective supervision, and transparency, and promote the transformation and upgrading of enterprises to achieve the goal of controlling greenhouse gas emissions. In the future, we will promote the construction work in three steps. The first step is the basic construction period which completes the national unified data reporting system, registration system and trading system construction in a year or so, and carry out the construction of the carbon market management system. The second step is to carry out the quota simulation transaction of the power generation industry, comprehensively test the effectiveness and reliability of various factors in the market, strengthen the market risk early warning and prevention and control mechanism. The last step is to deepen the improvement period which will carry out quotas among the trading entities in the power generation industry spot trading, under the premise of stable operation of the carbon market in the power generation industry, gradually expand the market coverage and enrich the trading varieties and trading methods.

3.9.3 Promoting South-South cooperation

The 2017 CCICED Policy Recommendations points out that the green "Belt and Road" concept and mechanism should also be reflected in South-South cooperation. Under the frameworks of South-South cooperation, "Belt and Road Initiative", and "BRICS plus", we should help other developing countries accelerate development through green transformation.

As one of the developing countries, China is an active advocate and supporter of South-South cooperation, and has always been combining its own interests with that of developing countries and providing assistance to South-South cooperation within its capacity. For more than 60 years, it has provided nearly 400 billion yuan of assistance for 166 countries, trained more than 12 million person-times of talents of various types for developing countries, and dispatched more than 600,000 assistance workers, of which more than 700 have sacrificed their lives for the development of other countries. It has set up the Assistance Fund for South-South Cooperation, the South-South Climate Cooperation Fund, the China-UN Peace and Development Fund, the Academy of South-South Cooperation and Development, the Center for International Knowledge on Development, and the South-South cooperation poverty reduction knowledge sharing website.

In January, 2018, in promoting the cooperation of the BRICS, the China Council for BRICS Think-tank Cooperation, Chinese coordinator of the BRICS think tank cooperation, held the 2018 Annual Meeting of the China Council and the Wanshou Forum. It has played an important role as a bridge and bond for promoting exchanges and cooperation of the BRICS think tanks and enhancing mutual understanding and trust.

At the beginning of September, 2018, the Beijing Summit of the Forum on China-

Africa Cooperation was held. The theme of this summit is "China and Africa: Toward an Even Stronger Community with a Shared Future through Win-Win Cooperation". The conference made two major achievements. First, it adopted the Beijing Declaration-Toward an Even Stronger China-Africa Community with a Shared Future, which pushed the China-Africa comprehensive strategic partnership to an even higher level, that is, a closer community with a shared destiny. Second, it adopted the Beijing Action Plan. The Plan covers the period from 2019 to 2021, and reflects the specific contents of cooperation between China and Africa in the next three years or even longer, especially the specific implementation of the "eight major initiatives" proposed by China. The Beijing Summit provided new ideas for South-South cooperation. It is noteworthy that in 2018, in a new round of institutional adjustment and reform, the China International Development Cooperation Agency was set up directly under the State Council to integrate the foreign aid duties of the Ministry of Commerce and the Ministry of Foreign Affairs. Its main responsibilities include formulating foreign aid strategic guidelines, plans, policies, coordinating major foreign aid issues and making recommendations, promoting foreign aid reform, preparing foreign aid programs and plans, determining foreign aid projects, and supervising and assessing their implementation. The specific implementation of foreign aid is still assigned to relevant departments. The establishment of the China International Development Cooperation Agency opened a new chapter in South-South cooperation for China.

As socialism with Chinese characteristics enters a new era, the relationship between China and the rest of the world is undergoing historic changes. Compared with any time in the past, China is more capable and confident to help developing countries improve their domestic development capabilities, shape the international development environment, improve mutual cooperation mechanisms, and inject greater vitality, create stronger momentum, and provide better pathways and broader space for South-South cooperation.

3.9.4 Participating in global ocean governance

According to the 2017 CCICED Policy Recommendations, China should formulate a national marine strategy to promote the development of the "Blue Economy" in a green direction. Due to the global character of China's Blue Economy, China can play an important role in the modernization of global ocean governance. The 2012 CCICED Policy Recommendations points out that a "national marine emergency response planning system for major environmental incidents" should be established, and relevant departments should co-develop emergency response plans for super and major marine environmental incidents as national marine emergency response special plans.

In February 2018, the State Oceanic Administration issued the National Marine Ecological Environment Protection Plan (2017—2020) (hereinafter referred to as Plan), systematically planning the timetable and roadmap for marine ecological environment protection in the following period of time. The Plan requires all relevant departments and units to take the implementation of the Plan as an important measure for implementing the spirit of the 19th National Congress of the Communist Party of China and deepening the construction of marine ecological civilization, refine the division of tasks, decompose responsibilities and objectives, clarify implementation pathways, and well guarantee organizational support so as to ensure the work items listed in the Plan render actual results.

The Plan identifies the principles of "green development, sea protection from the source" "compliance with nature, ecology-based sea management" "quality improvement, clean sea through cooperation" "reform and innovation, law-based sea governance" and "extensive mobilization, marine development through joint efforts". The marine ecological civilization system is basically complete, the marine ecological environment quality is steadily changing for the better, the green development level of marine economy is effectively improved, the marine environment monitoring and risk prevention and handling capacities are significantly improved. Goals are set for 4 aspects, and 8 indicators are proposed, such as the ratio of areas with good water quality in offshore areas, and the retention rate of land natural shorelines.

The Plan puts forward related work from the 6 aspects of "governance, use, protection, monitoring, control, and prevention", including promoting the governance and restoration of marine environment, carrying out systematic restoration and comprehensive governance in key areas, and improving the quality of marine ecological environment; establishing a green oceanic development pattern, and accelerating the establishment and improvement of a modern economic system with green, low-carbon and cyclic development; strengthening marine ecological protection, comprehensively safeguarding the stability of marine ecosystems and marine ecological functions, and building a marine ecological security barrier; adhering to "optimization of overall layout, strengthening of operation management, and improvement of overall capacity", and promoting marine ecological environment monitoring to enhance capacity and efficiency; strengthening land-sea pollution joint prevention and control, and implementing comprehensive prevention and control of pollution in river basin environment and coastal waters; preventing and controlling marine ecological environment risks, and building a whole-process and multi-level risk prevention system containing prevention, control, and handling before, during, and after the occurrence of risks.

In 2018, after the reform of state institutions,MEE will undertake and fulfill the functions of marine ecological environment protection, mainly including the supervision of the national marine ecological environment, the supervision of the discharge of land-sourced pollutants into oceans, the protection of ecological environment by controlling marine pollution and damage caused by coastal and marine construction projects, marine oil and gas exploration and development, and dumping of wastes into oceans, and the organization and demarcation of marine dumping areas.

On March 8th of 2018, the National Emergency Response Plan for Major Oil Spill at Sea (hereinafter referred to as Response Plan) was issued upon the review and approval by an inter-ministerial joint meeting for responding to national major oil spill at sea. The Response Plan established and improved the national major oil spill response procedures, and clarified the standards for national major oil spill incidents. It adheres to unified leadership, resource sharing, and efforts coordination, and reflects policy and guidance. After emergency response is initiated in a major oil spill, according to the Response Plan, the inter-ministerial joint meeting shall organize the implementation of national response measures, including: guiding the on-site command unit to formulate a scientific oil spill response plan and dispatch working groups to direct the on-site work; coordinating member units and other relevant forces in participating in oil spill response work such as monitoring and pollution removal; coordinating emergency response resources such as transportation, medical and health rescue, communication, emergency funds, technical equipment, human resources, and decision support; releasing or authorizing a relevant unit to release information on the oil spill, collecting and analyzing public opinions, and carrying out publicizing and reporting. Letting the inter-ministerial joint meeting play a commanding role in an emergency and coordinate the work of different departments and units can promote an orderly management of emergency resources.

3.10 Conclusions

In 2018, the Chinese government has made several greater breakthroughs in many aspects. In May of 2018, the National Ecological Environmental Protection

Conference established Xi Jinping's ecological civilization thought, emphasizing that a good ecological environment is the most inclusive benefit of the people, and that the lucid waters and lush mountains are invaluable assets.

China's eco-environment management system and mechanism have undergone major changes, and the reforms have been unprecedented. Eco-environment protection and supervision have achieved initial unification. At the same time, the high-level coordination mechanism at the State Council level has become more rational, and is moving toward a direction more conducive to unified supervision and coordinated development of ecological environment. China's eco-environment protection work has shown a good momentum and is actively moving toward the goals of 2035.

Over the past year, many policy recommendations put forward by CCICED continued to receive great attention. Many of them have been reflected in policy practice to different degrees. In particular, as the year 2018 is the initial period and critical period during the three-year battle on pollution prevention and control, China has successively issued several guidance documents to vigorously promote the implementation of eco-environment protection work, especially the prevention and control of air pollution. At the same time, the functions of environment protection departments were further optimized and expanded, the eco-environment supervision and governance mechanisms were rationalized, and the high-level coordination mechanism was smoother.

At present, all aspects of work are in a transitional period of coordination and rationalization. China's eco-environment protection work has been forging ahead, and the country continues to address outstanding environmental issues such as air, water, and soil pollution, advance environmental economic policies, promote public-private cooperation in environmental protection, and establish a pluralistic system of environmental governance. With the deepening of the ecological civilization system and the implementation of the Amendment to the Environmental Protection Law, all aspects of work have been promoted unprecedentedly. This will lay a good foundation for future environmental policy formulation and environmental governance improvement. At the same time, international environmental cooperation continues to show high-level development, including the cooperation on eco-environment between China and its neighboring Southeast Asian countries, and between China and African countries. Also, new layouts and work adjustments have been made on global concerns such as marine eco-environment protection.

Looking back at the environmental and development policies of the Chinese government in 2018, we can find that an institutional mechanism and a management system with "ecological civilization" at the core and "beautiful China and clean world" as the objective is rapidly forming. The overwhelming situation of China's eco-environment protection is helping China transform to the high-quality development goal. The top-level design and pluralistic governance system for China's ecological civilization have initially taken shape. China's ecological civilization and eco-environment protection are crucial to the realization of the United Nations 2030 SDGs. Its connection with the rest of the world regarding sustainable development issues has never been so close.

Appendix:

Overview on the Relevance of China's Environmental and Development Policies and CCICED Policy Recommendations over the Past Year

Field	Time of Release of Policy	Policy Progress (2017—2018)	Content
Top-level Design for Ecological Civilization	May 2017	The 8th National Eco-Environment Protection Conference put forward Xi Jinping's thought on ecological civilization, mainly including the profound historic view that "ecological development is civilization development", the scientific natural view of "harmony between man and nature", the green development view that "green mountain and clear water are valuable assets", the basic livelihood view that "a good ecological environment is the most inclusive benefit for people's livelihood", the overall systematic view that "mountains, waters, forests, crops, lakes, and grasses are a community of life", the strict legal view of "implementation of the most stringent eco-environment protection system", the all-people action view of "co-building a beautiful China", and the win-win global view of "co-planning global ecological civilization pathways"	According to the 2017 CCICED Policy Recommendations, China's voice and experiences in building an Ecological Civilization can be very powerful and inspirational for other countries. Domestically, it is important to build a strong linkage between SDGs and China's Five-Year Plans and to use the SDGs as a framework and basis for improving government policies and the efforts of business and communities.
Planning for Environment and Development	July 2017	The Ecological Environment Protection Plan for the Yangtze River Economic Belt: The Plan proposes to delineate the upper line for water resources utilization, delineate ecological protection red line, adhere to the bottom line of environmental quality, comprehensively promote environmental pollution control, and strengthen the prevention and response of sudden environmental incidents, making a comprehensive plan for the Yangtze River basin. As for implementation, the Plan puts forward an innovative measure of "innovating eco-environment protection mechanism policy and promoting regional coordination and connection".Relevant provinces and cities of the Yangtze River basin should firmly establish the concept of ecological community, strengthen overall, professional and coordinated regional cooperation, professional and coordinated, accelerate reform and innovation of institutional mechanisms, create a policy environment conducive to ecological priority and green development, and comprehensively enhance the level of coordinated protection of the ecological environment of the Yangtze River Economic Belt	According to the 2017 CCICED Policy Recommendations, fragmentation of decision-making remains a serious problem. At a fundamental level, integrated efforts for addressing land and water use on a regional scale are necessary, including better information sharing among regulatory agencies. Definition of responsibility for preservation and restoration of ecosystems can still be improved

Field	Time of Release of Policy	Policy Progress (2017—2018)	Content
Planning for Environment and Development	December 2017	Focusing on the development goals and strategies for 2035 and 2050, the National Energy Administration began to compile an energy development strategy outline that meets two phased goals. The initial design is as follows: First, for 2020—2035, actively promote market-oriented reform and institutional mechanism innovation, promote scale development and technological progress in the renewable energy industry, and promote the formation of development cost advantage of renewable energy over fossil energy. By 2035, the increase in China's energy demand can be met by clean energy, and renewable energy development enters an incremental replacement stage. Second, for 2035 until this mid-century, comprehensively build a modern energy system with renewable energy as the main part, and renewable energy enters the stage of full stock replacement of fossil energy. By 2050, its proportion in primary energy consumption reaches 60%, and its proportion in power consumption reaches 80%. Renewable energy becomes the main force of energy supply. Before this, we should ensure that China fully completes energy transformation	According to the 2017 CCICED Policy Recommendations, clean coal and synthetic natural gas for power generation should be transient technologies, bridging from old to new during China's green transition. Large-scale deployment of clean coal needs an exit plan and an exit budget to protect China from being locked into a path of prolonged fossil fuel use. Green adaptive planning should be part of the strategy
	April 2018	The Planning Outline for the Xiong'an New Area in Hebei Province proposes to build a scientific and rational spatial layout, create a beautiful natural ecological environment, improve industrial space layout, develop a green and intelligent new city, and build a high-standard green development featured Xiong'an New Area; adhere to green and low-carbon development, build a green municipal infrastructure system, promote eco-environment restoration, carry out environmental governance, build a green and intelligent transportation system, promote green and low-carbon production and lifestyle, and build a clean and environment-friendly heating system; scientifically use geothermal resources, coordinate natural gas, electricity, geotherm, biomass and other energy sources, and form a clean heating system with multiple energy sources; pay attention to the coordinated development of various systems, and take the Plan as the guidance, detailed control plans as the focus, and special plans as the support to form a planning system with universal coverage, hierarchical management, classified guidance, and integrated planning; carefully plan for each inch of land before starting construction, consider construction time sequence, deepen and refine detailed control plan, construction plan and various special plans so as to get ready for the comprehensive construction of the new area	The 2017 CCICED Policy Recommendations proposes to build a comprehensive eco-reform process for green development and ecological civilization. It states that today's rapid urbanization must become green urbanization, including ecologically sensitive planning with improved utilization of resources like space, soil and waste

Field	Time of Release of Policy	Policy Progress (2017—2018)	Content
Planning for Environment and Development	June 2018	The Opinions on Comprehensively Strengthening Ecological Environmental Protection and Securing a Decisive Victory in Pollution Prevention and Control proposes that by 2020, the overall quality of ecological environment shall be improved, the total discharge of major pollutants shall be greatly reduced, environmental risks shall be effectively controlled, and the level of eco-environment protection shall adapt to the goal of building a moderately prosperous society. At the same time, it proposes to accelerate the construction of an ecological civilization system to ensure the overall formation of a spatial layout, industrial structure, production mode, and lifestyle that conserve resources and protect ecological environment by 2035, the quality of the eco-environment is fundamentally improved, and the goal of building a beautiful China is basically realized. By the middle of this century, ecological civilization is comprehensively upgraded, and modern national governance system and capability in the eco-environmental field are achieved	According to the 2017 CCICED Policy Recommendations, while the Air Action Plan is the first to come up for renewal, it would be helpful to develop an overall plan that incorporates all three categories (water, air, and soil), plus one other-marine pollution. Ideally, an integrated rollout should be ready by 2020 with targets up to 2035, the pivot point when China expects to be a "basic modern country"
		The Three-year Action Plan for Winning the Blue Sky War proposes to, with 3 years of efforts, greatly reduce the total emissions of major air pollutants, work together to reduce greenhouse gas emissions, further significantly reduce the concentration $PM_{2.5}$ and heavy pollution days, greatly improve ambient air quality, and greatly enhance people's sense of happiness with the blue sky. By 2020, the total emissions of sulfur dioxide and nitrogen oxides are reduced by more than 15% compared with 2015; cities of and above the prefecture level with excess $PM_{2.5}$ concentrations shall cut the concentrations by over 18% compared to 2015, the proportion of days with superior air quality in cities of and above the prefecture level shall reach 80%, and the proportion of days with severe pollution and above shall decrease by over 25% compared to 2015; provinces that have completed the targets and tasks for the 13th Five-Year Plan period shall maintain and consolidate the improvement results; provinces that have not completed them shall ensure the full realization of the binding targets for the 13th Five-Year Plan period; Beijing's ambient air quality improvement target should be further enhanced on the basis of the 13th Five-Year Plan period	

Field	Time of Release of Policy	Policy Progress (2017—2018)	Content
Ecosystem and Biodiversity Conservation	2017—2018	Positive progress has been made in the delineation and protection of ecological red lines. Beijing City, Tianjin City, and Hebei Province, the 11 provinces (cities) in the Yangtze River Economic Belt, and the Ningxia Hui Autonomous Region, a total of 15 provinces, have had their ecological protection red line delineation plans approved by the State Council, and Shanxi Province and the other 15 provinces have initially formed their delineation plans. The construction of a national ecological protection red line supervision platform is started. In October 2017, the National Development and Reform Commission officially approved the project on national ecological protection red line supervision platform with a total investment of 286 million yuan. Study and formulate supporting management policies for ecological protection red line. Issue guidance documents such as the Guidelines for Delineating Ecological Protection Red Lines. Study and draft ecological protection red line management measures, and clarify the control requirements, management principles and regulatory framework for ecological protection red line. At present, a draft for comments has been formed, and relevant departments and local governments will be consulted	The 2014 CCICED Policy Recommendations proposes to implement a National Ecological Protection Red Line System (EPRL System): set into law the National Ecological Protection Red Line (EPRL) System and relevant systems; improve spatial land use planning and marine use planning system with clear identification of EPRLs; establish a new national coordinating mechanism for ecological conservation and for monitoring and enforcement; improve the nature protection area system; improve eco-compensation and incentive mechanism based on EPRLs
	March 2018	The "Green Shield 2018" nature reserve supervision and inspection special action. Retrospect and review the rectification of problems in the "Green Shield 2017" special action; resolutely investigate and deal with new violations in nature reserves, focus on inspecting the inadequate implementation of management responsibilities in state-level nature reserves; strictly supervise problem identification and rectification in nature reserves	
	June 2018	According to the auditing results released by the National Audit Office on eco-environment protection of the Yangtze River Economic Belt, in 2016 and 2017, a total of 251.824 billion yuan of financial funding was invested in the eco-environment protection of the Economic Belt. While eco-environment protection has achieved certain results, there are also some problems. Among them: First, in terms of the use of funding, from December 2013 to January 2018, 8 local government departments and affiliated units illegally used 25.8049 million yuan of eco-environment protection fund. Second, in terms of resource development, as of the end of 2017, 930 small hydropower stations in 8 provinces started construction without environmental impact assessment. Over-exploitation caused 333 rivers to be cut off, with a total cutoff length of 1,017 kilometers. Third, in terms of pollution control, as of the end of 2017, 118 urban sewage treatment plants in sensitive areas of 9 provinces did not meet the class 1A discharge standards	

Field	Time of Release of Policy	Policy Progress (2017—2018)	Content
Energy, Environment and Climate	2018	Energy structure optimization and adjustment. A green and diversified energy supply system was accelerated. In the first half of the year, hydropower, nuclear power, wind power and solar energy accounted for 25.2% of the total power generation, an increase of 0.3 percentage points over the same period last year. In terms of coal, adopt the principle of replacement at a reduced amount to develop high-quality coal production capacity in an orderly manner, force inefficient and low-quality production capacity to exit as soon, and increase the proportion of advanced coal production capacity. The power consumption structure was further optimized. In the first half of the year, the proportion of electricity consumption by the secondary industry was 69.2%, down by 1.8 percentage points from the same period of last year. The proportion of total electricity consumption by the four energy-intensive industries was 28.5%, down by 1.3 percentage points over the same period of last year; the proportion of electricity consumption by the tertiary industry and by households was 29.8% in total, up by 2.6 percentage points over the same period of last year. The contribution rate of secondary industry electricity use to the growth of electricity use by the whole society was 56.5%, which was 12.2 percentage points lower than the same period of last year; the contribution rate of tertiary industry electricity use and residential electricity use to the whole society's electricity consumption growth was 42.4% in total, up by 13.0 percentage points compared with the same period of last year. The power consumption structure underwent constant optimization	The 2017 CCICED Policy Recommendations states that through co-benefits, China's pollution reduction plans can contribute to a steady transition for meeting the Paris targets of staying within a global 1.5 or 2℃ increase
	May 2018	Strengthen energy conservation and improve energy efficiency. Revise the Administrative Measures for Energy Conservation in Key Energy-using Units, and establish and improve joint incentive system for integrity and joint punishing system for loss of integrity so as to accelerate social integrity building	The 2017 CCICED Policy Recommendations states that through co-benefits, China's pollution reduction plans can contribute to a steady transition for meeting the Paris targets of staying within a global 1.5 or 2℃ increase.

Field	Time of Release of Policy	Policy Progress (2017—2018)	Content
Energy, Environment and Climate	2018	Coordinate national efforts to respond to and adapt to climate change. Establish a National Leading Group to Address Climate Change, Energy Conservation and Emission Reduction headed by the Premier of the State Council, including main officers from more than 20 ministries and commissions as group members. All provinces, municipalities and autonomous regions have set up corresponding climate change leading groups and mechanisms for work division and coordination among departments. A climate change response system is formed, with coordination work done by the National Leading Group and implementation work headed by competent departments in collaboration	The 2017 CCICED Policy Recommendations states that through co-benefits, China's pollution reduction plans can contribute to a steady transition for meeting the Paris targets of staying within a global 1.5 or 2℃ increase
	2018	Continue to explore the construction of a national carbon market. Strengthen the carbon market management system, infrastructure and capacity building. Promote the issuance of an Interim Regulations on the Management of National Carbon Emissions Trading and related supporting systems such as quota allocation. Accelerate the construction of the data reporting system, registration system, trading system, and settlement system. Organize capacity building activities for various market players	The 2017 CCICED Policy Recommendations states that through co-benefits, China's pollution reduction plans can contribute to a steady transition for meeting the Paris targets of staying within a global 1.5 or 2℃ increase
Pollution Prevention and Control	2018	Air pollution prevention and control. The Ministry of Ecology and Environment implemented two rounds of intensive supervision of the "2+26" cities in the Beijing-Tianjin-Hebei region and surrounding areas. Despite remarkable results in air pollution control, 5,204 atmospheric environment problems were identified	According to the 2013 CCICED Policy Recommendations, concerning implementation of the Air Pollution Control Action Plan, the central government should focus its supervision and coordination efforts on strengthening overall action implementation by local governments and step up review and accountability

Field	Time of Release of Policy	Policy Progress (2017—2018)	Content
Pollution Prevention and Control	Throughout 2018	Water pollution prevention and control. The Ministry of Ecology and Environment and the Ministry of Housing and Urban-Rural Development launched a special rectification action for black and odorous water bodies, and initiated inspections. Inadequate rectification of identified problems will be held accountable by the Central Environmental Inspection Group. The water body in drinking water source regions will rank the first among the "four types of water bodies" in water pollution prevention and control. The Ministry of Ecology and Environment and the Ministry of Water Resources have launched a special action on environmental protection of centralized drinking water sources throughout the country	The 2015 CCICED Policy Recommendations proposes to establish an environmental risk assessment and prevention system for major national strategies, conduct risk assessments for macro strategies such as the Belt and Road Initiative, the integration of the Beijing-Tianjin-Hebei region, and the development of the Yangtze River economic belt to form an integrated environmental risk prevention system. The Notice on Guiding Opinions on Strengthening Yangtze River Environmental Pollution Prevention and Control puts the restoration of the eco-environment of the river at a dominant position. With water environment quality improvement at the core, it will strengthen spatial management, optimize industrial structure, consolidate pollution control from the source, and pay attention to risk prevention so as to fully advance Yangtze River water pollution control and ecological protection and restoration

Field	Time of Release of Policy	Policy Progress (2017—2018)	Content
Pollution Prevention and Control	2018	Soil pollution prevention and control. The Ministry of Ecology and Environment comprehensively promoted the 10-Chapter Soil Pollution Action Plan, and formulated the Regulations on the Assessment of the Implementation of the Soil Pollution Prevention and Control Action Plan, prescribing the annual assessment and mid-term assessment of the implementation of the Action Plan by provinces, municipalities, and autonomous regions from 2018 to 2020	Regardless of promoting ecological civilization or building the relationship between green and harmonious environment and social development, the Chinese government must well address outstanding issues affecting public health and living, including air, water and soil pollution and the decrease of ecological functions
Environmental Governance and Rule of Law	March 2018	The newly formed Ministry of Ecology and Environment will serve as a department of the State Council, integrating the responsibilities of the Ministry of Environmental Protection, the responsibilities of the National Development and Reform Commission for addressing climate change and emission reduction, that of the Ministry of Land and Resources for supervising and preventing groundwater pollution, that of the Ministry of Water Resources for preparing water function zoning, managing drainage outlets settings, and protecting watershed water environment, that of the Ministry of Agriculture for supervising and guiding agricultural non-point source pollution control, that of the State Oceanic Administration for marine environmental protection, and that for environmental protection in the project area of the South-to-North Water Diversion Project under the Office of the State Council's South-to-North Water Diversion Project Construction Committee	Article 2 of the 2017 CCICED Policy Recommendations proposes to "build Ecological Civilization co-management systems". The 2013 CCICED Policy Recommendations proposes to speed up institutional reform for eco-environmental protection management; establish an environmental governance system for unified supervision of all pollutants, all emission sources, all environmental components, and all ecosystems

Field	Time of Release of Policy	Policy Progress (2017—2018)	Content
Environmental Governance and Rule of Law	Early 2018	The Ministry of Ecology and Environment promulgated the Interim Measures for Pollution Permits in early 2018, and a classified management list of pollution permits will be released. It Issued the general rules for the thermal power and paper making industries, such as the Technical Specification for the Application and Issuance of Pollution Permits for Thermal Power and Paper Industry and the Self-monitoring Technology Guidelines for Pollution Sources, and initially established a relatively complete system of regulations and technical specifications to fully support the application and issuance of pollution permits. A national unified pollution permit management information platform has been basically built and put into operation. The permit application and issuance of the two industries of thermal power and paper making are all carried out on the information platform, basically achieving a simple, accurate and efficient information management mode	The 2015 CCICED Policy Recommendations proposes to draft an Emission Permits Law to integrate the emission permit system within the broad system of environmental standards, environmental monitoring, environmental impact assessment, integration of the concept known as "Three Simultaneous", namely pollution emission registration, total emission control, and the regulation of environmental facilities and management of emission discharge outlets; enhance the legal status of the emission permit system, and ensure this system forms the core of environmental management
	May 2018	On May 7, 2018, the Ministry of Ecology and Environment, in principle, adopted the Administrative Measures for Compulsory Liability Insurance for Environmental Pollution and established an "environmental pollution compulsory liability insurance system" in areas with high environmental risks. The People's Bank of China and the China Securities Regulatory Commission jointly issued guidelines to standardize the evaluation and certification of green bonds	The 2015 CCICED Policy Recommendations proposes to promote green credit and develop a market for green bonds and green insurance through innovative means; in high environmental risk areas, implement a compulsory environmental liability insurance system; support and encourage financial institutions and enterprises to issue green bonds
	First half of 2018	The People's Bank of China and the China Securities Regulatory Commission jointly issued guidelines to standardize the evaluation and certification of green bonds	
	June 2018	The China Securities Regulatory Commission issued an announcement, proposing to seriously rectify the illegal behavior of listed companies in disclosing environmental information, continue to maintain the high pressure of law enforcement on major environmental pollution information disclosure violations, comprehensively and strictly enforce administrative penalties in accordance with the law, urge listed companies to effectively fulfill their eco-environment protection obligations, guide them to practice their corporate social responsibility, and strive to secure a decisive victory in capital market pollution prevention and control and in the construction of ecological civilization for a long period of time	

Field	Time of Release of Policy	Policy Progress (2017—2018)	Content
Environmental Governance and Rule of Law	June 2018	The National Development and Reform Commission issued the Opinions on Innovating and Improving a Price Mechanism Promoting Green Development in June 2018. The Opinions points out that given the new situation and new requirements of ecological civilization construction and eco-environment protection in a new era, it is necessary to fully utilize market-based means to promote the marketization of eco-environment protection, continuously improve the resource and environment price mechanism, and better utilize the positive functions of price lever in guiding the rational allocation of resources, realizing the internalization of eco-environmental costs, promoting whole-society conservation, and accelerating the development of green and environmental protection industries, thus stimulating the whole society to jointly promote green development and ecological civilization. The Opinions proposes that by 2020, a price mechanism and price policy system conducive to green development shall be basically formed, with a significantly enhanced role in promoting resource conservation and internalization of eco-environmental costs; by 2025, a price mechanism that meets the requirements of green development shall be more complete and implemented in all aspects of the whole society	The 2015 CCICED Policy Recommendations proposes to reform the pricing mechanism for critical resources, and use fossil fuels (such as coal and oil) as entry points to internalize environmental costs into a pricing mechanism; develop green financial and taxation policies to reflect environmental costs of production and consumption; create a market environment with healthy competition for green industries and actively promote energy-saving and environmental industries
	2018	In 2018, the judicial department issued the latest full text of the General Rules of the Civil Law. Regarding the provisions on principles, article 9 proposes that civil subjects engaged in civil activities shall facilitate resource conservation and eco-environment protection. This is the first principle in the general rules of the civil law to require consistency between civil acts and resource conservation and eco-environment protection, and has become a green principle. This is a close echo with the formation of green lifestyles proposed at the 19th National Congress of the Communist Party of China. On June 4, 2018, the Supreme People's Court issued the Opinions on Studying and Implementing Xi Jinping's Thought on Ecological Civilization to Provide Judicial Service and Guarantee for Ecological Environment Protection in the New Era, requiring courts at all levels to give better play to their judicial functions in environment and resources and strengthen judicial service and guarantee for ecological civilization	The 2014 CCICED Policy Recommendations proposes to improve environment and health related institutions; include environmental health risk assessment in the making of polices and standards; improve the environmental public interest litigation system and the ecological environment damage compensation and accountability systems; strengthen responsibility and capacity of the judicial system to investigate environmental violations that result in injury to people

Field	Time of Release of Policy	Policy Progress (2017—2018)	Content
Regional and International Engagement	2018	Integrating the concept of green development into the Belt and Road Initiative, and promoting Belt and Road projects to bring about sustainable development effects both economically and environmentally have provided new impetus for the implementation of the 2030 Sustainable Development Goals. At present, 15 United Nations agencies and relevant Chinese institutions have signed agreements related to the Belt and Road Initiative. The core of these agreements is to create synergies between the 2030 Agenda for Sustainable Development and the green Belt and Road Initiative	Article 5 of the 2017 CCICED Policy Recommendations states as follows: China's green development approach, the UN 2030 SDGs, the Paris Agreement, Biodiversity and Desertification global accords, and ecological civilization should become central features in the Belt and Road Initiative...Also, a Belt and Road Green Governance Mechanism including information disclosure, public participation and access to arbitration, should be established
	2018	Since the beginning of 2018, the focus of the work on carbon market has shifted from pilot demonstration to building a unified national market. Carbon trading pilot areas should further deepen pilot work, improve the design of the pilot carbon market system, summarize piloting experience, and gradually transit to a national market after conditions are mature, on the basis of maintaining the stable operation of the pilot carbon market	Article 5 of the 2017 CCICED Policy Recommendations states as follows: Strengthen global and regional green governance. It proposes that "China needs to start early to develop its own Mid-Century Climate Strategy and link it with others"

Field	Time of Release of Policy	Policy Progress (2017—2018)	Content
Regional and International Engagement	March 2018	The China International Development Cooperation Agency was set up directly under the State Council to integrate the duties of the Ministry of Commerce for foreign aid and the duties of the Ministry of Foreign Affairs for foreign aid coordination. Its main responsibilities include formulating foreign aid strategic guidelines, plans, policies, coordinating major foreign aid issues and making recommendations, promoting foreign aid reform, preparing foreign aid programs and plans, determining foreign aid projects, and supervising and assessing their implementation. The specific implementation of foreign aid is still assigned to relevant departments. The establishment of the China International Development Cooperation Agency opened a new chapter in South-South cooperation for China	According to the 2017 CCICED Policy Recommendations, the green Belt and Road concept and mechanism should also be reflected in South-South cooperation. Under the frameworks of South-South cooperation, Belt and Road Initiative, and "BRICS plus", we should help other developing countries accelerate development through green transformation
	September 2018	The Beijing Summit of the Forum on China-Africa Cooperation made two major achievements. First, it adopted the Beijing Declaration-Toward an Even Stronger China-Africa Community with a Shared Future, which pushed the China-Africa comprehensive strategic partnership to an even higher level, that is, a closer community with a shared destiny. Second, it adopted the Beijing Action Plan. The Plan covers the period from 2019 to 2021, and reflects the specific contents of cooperation between China and Africa in the next three years or even longer, especially the specific implementation of the "eight major initiatives" proposed by China. The "eight major initiatives" cover a variety of contents, which enables China-Africa cooperation to build on the existing foundation and achieve a higher strategic level. The Beijing Summit provided new ideas for South-South cooperation	
	February 2018	The State Oceanic Administration issued the National Marine Ecological Environment Protection Plan (2017—2020) (hereinafter referred to as Plan), systematically planning the timetable and roadmap for marine ecological environment protection in the following period of time. The Plan requires all relevant departments and units to take the implementation of the Plan as an important measure for implementing the spirit of the 19th National Congress of the Communist Party of China and deepening the construction of marine ecological civilization, refine the division of tasks, decompose responsibilities and objectives, clarify implementation pathways, and well guarantee organizational support so as to ensure the work items listed in the Plan render actual results	According to the 2017 CCICED Policy Recommendations, China should formulate a national marine strategy to promote the development of the "Blue Economy" in a green direction. Due to the global character of China's Blue Economy, China can play an important role in the modernization of global ocean governance.

Field	Time of Release of Policy	Policy Progress (2017—2018)	Content
Regional and International Engagement	March 2018	The National Emergency Response Plan for Major Oil Spill at Sea (hereinafter referred to as Response Plan) was issued upon the review and approval by an inter-ministerial joint meeting for responding to national major oil spill at sea. The Response Plan established and improved the national major oil spill response procedures, and clarified the standards for national major oil spill incidents. It adheres to unified leadership, resource sharing, and efforts coordination, and reflects policy and guidance. After emergency response is initiated in a major oil spill, according to the Response Plan, the inter-ministerial joint meeting shall organize the implementation of national response measures, including: guiding the on-site command unit to formulate a scientific oil spill response plan and dispatch working groups to direct the on-site work; coordinating member units and other relevant forces in participating in oil spill response work such as monitoring and pollution removal; coordinating emergency response resources such as transportation, medical and health rescue, communication, emergency funds, technical equipment, human resources, and decision support; releasing or authorizing a relevant unit to release information on the oil spill, collecting and analyzing public opinions, and carrying out publicizing and reporting. Letting the inter-ministerial joint meeting play a commanding role in an emergency and coordinate the work of different departments and units can promote an orderly management of emergency resources	According to the 2012 CCICED Policy Recommendations, a "national marine emergency response planning system for major environmental incidents" should be established, and relevant departments should co-develop emergency response plans for super and major marine environmental incidents as national marine emergency response special plans.

Chapter 4

Shocks, Innovation and Ecological Civilization A "New Green Era" for China and for the World

4.1 Introduction

In 1970 Alvin Toffler published a "must read" book, which described the evolution towards post-industrial, information-oriented societies characterized by impermanence, and fear of the future.[1]Future Shock was written at the dawn of a global "new era" initially dominated by wealthier nations that had new means of communication, almost unlimited mobility, and innovation led by the Bretton Woods institutions that set the stage for globalization of trade and investment. Particularly in the 1980s and 1990s, market forces ruled in the western world. Over time development benefits widened to include many developing countries, with poverty reduction for some, and dramatic economic growth in the case of China.

It can be argued that sunset for this particular "new era" period was the 2007—2008 global financial crisis,[2] which exposed many governance issues, and created widespread antipathy towards excesses associated with globalization, and towards inequalities in wealth creation and distribution. Often this antipathy is expressed as rejection of expert opinion of "elites", especially on the part of those who feel "left behind" in post-industrial societies. Today we see the rise of populism in many countries, and calls for major institutional changes—including a shift away from many of the international economic agreements that have shaped social and economic progress during the last 40 to 50 years. This has resulted in both political and economic shocks in nations as diverse as the USA, Brazil and Venezuela, the UK and some other EU members, and parts of Africa.

Coincident with the start of the 1970's "new era" was the emergence of environment and development as a major global concern. From the 1972 Stockholm Environment Conference to the UN 2030 SDGs Framework and the 2015 Paris Agreement, there has been a steady rise in concern for environment globally and at national levels. This rise has led to innovation technologies, planning and financing progress, and growing participation by business, communities and citizens. Despite this positive effort the threats to the planet's ecosystems and environmental protection have continued to become more acute and self-reinforcing.[3]

Now, in various parts of the world hard-won environmental progress is under considerable threat as a consequence of several factors including population growth, rising per capita consumption for wealthier countries, and extreme poverty.

1 ALVIN TOFFLER. Future Shock. Bantam. 1984.
2 MARTIN WOLF. The Shifts and the Shocks. Penguin. 2015.
3 The most recent warning was from the IPCC in October 2018 indicating that unless global temperature rise was maintained below 1.5℃,planetary conditions for humans and natural ecosystems will face great difficulties. https://www.ipcc.ch/pdf/session48/pr_181008_P48_spm_en.pdf.

A new political pressure is the rise of populism as seen in the USA and an alarming number of other nations. This pressure is sometimes accompanied by a rejection of science, globalism, and governmental interventions such as green taxes.

The need for transformative changes to environment and development relationships cannot be denied. We know what is needed and the urgency of creating successful interventions. Indeed, through innovation there are many economic opportunities to be harvested through green development. Still, there will be shocks ahead. A bumpy ride can be predicted.

It is time for a "new green era" to take hold—where environment and development issues are firmly linked and mainstreamed. We have the makings already in hand globally in the form of coherent, integrated development approaches via the UN 2030 SDGs framework, plus various global conventions, and environmental, economic and social institutions linked to the UN.

China's efforts nationally are supported by theory and practices for construction of an Ecological Civilization linking five key policy action areas: economy, social, cultural, environmental and political.[1] The EU has created an impressive policy effort involving its member nations.[2] However there is no guarantee of success, and even if goals are met, an ecologically and environmentally secure future will require much additional action.[3] The decade ahead is significant, since option foreclosure is already occurring, and by 2030 unfulfilled tasks will be much more difficult, especially for water use, green urbanization, biodiversity conservation, climate change mitigation and adaptation, and sustainable ocean use.

4.2 China's New Era[4]

China has signaled at the highest level that the country has entered its own New Era. It is a time of definitive change with signposts: 2020—a moderately prosperous society; 2035—basic realization of socialist modernization; and 2050—

1 XIE ZHENHUA ,PAN JIAHUA. China' s Road of Green Development. Beijing: Foreign Languages Press. 2018.

2 For example, the EU Roadmap for a low-carbon economy by 2050; the environmental action program vision to 2050 on innovation, circular economy and sustainability. https://ec.europa.eu/info/energy-climate-change-environment/overall-targets/2050-targets_en

3 PBL. The worldwide context of China' s green transition to 2050. PBL Publication 2982. 58. 2017.

4 For a useful overview see: China' s "New Era" with Xi Jinping Characteristics" . 15 December 2017. European Council on Foreign Relations. China Analysis. 16. https://www.ecfr.eu/page//ECFR240_China_Analysis_Party_Congress_Ideology

a prosperous, harmonious and beautiful country. This New Era is focused on the contradiction "between unbalanced and inadequate development and the people's ever-growing needs for a better life." Emphasis is placed on the achievement of an Ecological Civilization by 2035. While these points largely focus on domestic action so people can pursue a better life, there is also much about the New Era that is focused at a global level. China is indicating that it is prepared to tackle larger roles for strengthening global governance.

The notion of a "community with a shared destiny for mankind" links China and the World, with commitments to take on roles in reform and development of global governance. Recent statements have committed China to greening of the BRI, firm support for the Paris Agreement, commitment to a China 2030 SDGs Action Plan, hosting of the Convention on Biological Diversity (CBD) 2020 COP 15 (Convention of the Parties 15) in China that will set 2020—2030 goals for biodiversity; and other commitments that demonstrate willingness to take on some leadership roles internationally, and in general, to become a more active player on global environmental stewardship. It is developing greater confidence in its own development model and its innovation capacity to achieve economic and some social goals. This confidence may make it easier to transfer this experience to others.

China's New Era terminology arose from the 19th National Congress of the Communist Party of China held in October 2017, with subsequent March 2018 National People's Congress follow-up. It is largely based on President Xi Jinping's efforts of the past five years and future plans of the Party and Government under the leader's direction. The New Era signals a major national shift that will be played out in a global context where China is emerging as a strong player; and where other sorts of "rejuvenation" and "dismantlement" by some other countries or alliances are being proposed or undertaken.

Some of the most concrete examples in 2017 and 2018 regarding Environment and Development in China's New Era have been institutional reforms intended to make resource and environmental decisions within China more efficient and effective. Also, reform of the country's international development system; guidelines for greening of the financial sector; the Green Tax Law implementation; and other legislative reforms. Punitive action against those Party Members and civil servants not obeying environmental rules, or engaging in corrupt practices has become severe. The same is true for enterprises. Environment is now ranked as one of the highest three priorities for government action.[1]

If China is successful in meeting its New Era aspirations, the consequences will be very positive not only for China, but also for many other countries, especially some developing nations. There should be benefits regarding global environmental conditions as well. There are some international expectations that China can play a leadership role on climate change, on desertification and on biodiversity conservation. In the case of oceans, China is a very important player in world fisheries and aquaculture, shipping, port development, the blue economy and biodiversity matters including migratory species protection, and regulation of trade in endangered species, and in the international shipment of wastes. China is heavily engaged in ocean science and technology, including arctic and Antarctic safety. China can become a major contributor to ocean use monitoring. These examples suggest China can underpin the effort for sustainable development globally. In doing so, China may even find broad support among other countries for its ecological civilization approach.

4.3 Global Green New Era?

What of the bigger picture for a Global Green New Era on Environment and Development? Will the 2020s become the decade of environmental improvement so badly needed at a global level? Or will goals of important initiatives such as the Paris Agreement and the UN SDGs be missed? Will global enterprises soften their commitments to environmental quality if there is not sufficient external pressure? Can hard-pressed cities throughout the world keep up with their low carbon and other green goals—and still provide for rapid urbanization rates?

4.3.1 Environmental Risks

Environmental risks are on the rise. And they interact with other risks in complex webs that can affect social instability, disease incidence and many other types of risks. The important point is that even as environmental issues become more important risk factors, their potential and real impacts often are being downplayed for self-serving reasons. Also, there is some degree of resignation—for example, the rise in occurrence and intensity of forest fires in various areas of the world is being

1 A very comprehensive overview of Government of China action on environment is provided in 2017—2018 Progress on Environment and Development Policies in China and Impact of CCICED's Policy Recommendations. This document is prepared annually for distribution at the CCICED Annual General Meeting.

considered as a "new normal" as a consequence of climate change. Yet this may be misleading since it suggests a plateau in impacts when actually such situations can become worse and worse over time.[1]

Clearly, better environmental risk management at a global level is essential as part of a New Era where ecological safety, pollution control and environmental monitoring must prevail as essential conditions for staying within nine key planetary boundaries. The World Economic Forum (WEF) and others point to climate change as a major emerging risk to economic and other interests. This has been an important consideration for some time, as a consequence of rising claim levels from "natural disasters" such as floods. Very likely, the international re-insurance industry will become an essential driver for improved performance, especially for responsible environmental care and other programs involving financing. If enterprises in any sector are refused insurance, behavioral change takes place. Future shocks that potentially affect communities, ecosystems and economies should be a strong incentive for preventive action; and to some extent such efforts are taking place throughout the world.

Yet, as noted in the 2017 global risk assessment publication prepared for the WEF:

Global cooperation is under strain. Heightened domestic anxiety has intensified geostrategic competition. Advanced economies look to strengthen border controls, while climate change reform hovers under a cloud of uncertainty and trade agreements are collapsing.[2]

4.3.2 "Push and pull"

Unfortunately, it is becoming clear that some of the future shocks likely to affect green global progress will likely to be driven by ideological, political or some poorly-founded grounds rather than acting on gathered scientific evidence of the urgent need for global and national environmental protection and mitigation. This situation is becoming apparent in statements and action on climate change, on biodiversity where international success is highly dependent of the quality of national action, and on trade and investment—where even the World Trade Organization (WTO) appears to be under a degree of threat.

At present there is both push and pull globally. Much of the environmental push is driven by non-governmental organizations from both north and south countries,

1 https://www.theglobeandmail.com/politics/article-a-summer-of-fire-heat-and-flood-puts-a-focus-on-adapting-to-climate

2 https://www.marsh.com/content/dam/marsh/Documents/PDF/US-en/The%20Global%20Risks%20Report%202017-01-2017.pdf

plus the well organized associations and research bodies that have been of great value in the international negotiations. Since the 2012 Rio+20 Earth Summit and follow-up for the UN 2030 SDGs effort and the 2015 Paris Agreement negotiations, there has been an air of optimism that a gateway to rapid progress is opening. Even with the US pull away from the Paris Agreement, and various other disappointments such as limited action on greening of trade and investment agreements, there is still hope that accelerated progress can be made between now and 2020, when further assessments will take place.

4.3.3 China and a global green New Era

What are the implications for China's New Era and for Ecological Civilization in this turbulent global period? As always, there is opportunity in such times. First, by staying the course of innovation; second, by firm commitment to a forward-looking agenda of environmental improvement and investment in green development; and third, by enhanced participation in global green governance, China is already signaling its intent to each of these positions. And by committing to a green BRI, South-South cooperation, and greening of international financing bodies such as the Asian Infrastructure Investment Bank(AIIB), China's ability to work with developing nations on environment and development concerns will be strengthened. Such efforts will be particularly valuable if they can be linked to progress on the 2030 SDGs, and if countries are interested, promotion of ecological civilization.

China can be an important element in the push side of the equation. It will become more apparent over time that there are international competitive advantages to be had by pursuing green paths. This was demonstrated by Germany decades ago during standard setting for reusable and recyclable bottles. And by Denmark as it promoted wind power in the 1990s. China is betting on electric automobiles, advanced battery technology, and undoubtedly many other products and approaches being promoted for low carbon cities. China can expand its green approaches to food production, advanced circular economies, and innovative green practices for shipping and other forms of transportation, and in many sectors such as ecotourism, major sports events, advanced water conservation, and green industrial practices for both large and small enterprises.

Potentially, China's biggest asset for promoting significant attention to a new era for green global action is its huge population—now 1.4 billion, about 18.5 % of the world's total. The shift from an export driven to a domestic consumption oriented economy offers many opportunities for decoupling future economic growth from environmentally damaging styles of consumption. China's ecological footprint has grown significantly over recent times. However, on a per capita basis, today's Chinese footprint is still well below what is found in major cities outside of China and in rich countries, especially in North America. The coming 5 to 10 years will be a deciding factor on how effectively incentives can shape consumption patterns towards greener, low carbon, circular economy paths. This is especially the case for food consumption, since China is increasingly dependent on imported sources to satisfy growing demand for meat and fish. Chinese consumers equate clean environment criteria with supply of high quality "pure" food. These are examples where green supply chains and standards become very important.

As China transforms into an ecological civilization domestically, there will be many ramifications for trade and investment both internally and abroad. Intermediary organizations such as those setting production standards and certification can help companies to secure markets. Enterprises selling green technologies, sources providing access to green financing, and IT organizations can all expect to benefit. Job creation through green development is an obsession, whether within China, or other countries. Transformative change toward ecological civilization will need to demonstrate employment opportunities that exceed those otherwise available (for example, those in fossil fuel industries) in order to popularize this "Made in China" approach with wider audiences.

China can help move the goalposts for at least some of the 2030 SDGs initiatives by outperforming anticipated levels of progress. This approach was very helpful in securing the global success of the Millennium Development Goals (MDGs). By demonstrating rapid progress at an early stage, China not only showed the goals to be feasible, but also helped stimulate progress by others. We already know that even if 2030 SDGs are fully met by 2030, it is not enough to secure a sustainable future. To the extent that the pace of change can be accelerated in larger countries, time can be bought for the future.

Certainly, however, China cannot be expected to shoulder too heavy of a burden for either itself domestically, or on behalf of the world community. China has been quite explicit on this subject. This point is relevant to various reforms needed for global green governance. Means must be found to enhance the performance of international accords and to ensure that once agreed upon, they have a high degree of certainty regarding implementation. Also greater efficiency is required during implementation. At the June 2018 CCICED Brussels Roundtable various synergies were identified among the major global Conventions for Climate Change, Biological Diversity, and Oceans. Such an approach is not unique, but often the results have been paid only lip service. It is time to take full advantage of synergies in order to amplify outcomes. This is a topic that China could promote globally, and in its own initiatives.

It is fair to describe China as an emerging torchbearer on the international environment and development front. It can serve as a mediator between G7 and the G77. This role will be served well if backed up by leading through example, leading through strategic allocation of its financial resources and growing bank of science and technology and expertise, and leading with various coalitions for improving selected aspects of global green governance. In a world increasingly affected by the turmoil of change, China stands out as a nation equipped to be resilient and adaptive. These qualities are important at a time when polyvalent governance models are called for in order to provide flexibility and inclusiveness in addressing problems.

4.4 Innovation, Idealism and Pragmatism for a Sustainable Future

In summary, late 2018 brings the world to a brink characterized by major shifts including ecological local-to-global crises, deconstruction of important multilateral agreements, trade wars, post-truth societies, civil conflict and war with hardening of attitudes towards migration or even basic life support, and polarized views that are hollowing out centrist governments in many countries. Demographic trends include aging populations in some places, remaining areas of extreme and relative poverty, situations of extreme wealth concentration and environmentally unsustainable consumption, and rising levels of urbanization. There are unprecedented challenges and opportunities associated with the rapid rise of new technologies: the digital age including artificial intelligence, biotechnology and nanotechnology.

It is no wonder that shifts and future shock are a source of fear. Also that a sense of dystopia grips many people who see loss of work, loss of tradition and society as they know it, and in some cases corruption and mismanagement in governments at a totally unacceptable level. Yet there is also a sense of excitement and hope that drive many people in many locations worldwide. This hope exists not only for some who have prospered greatly in commerce or in professions, but also among those who may live modestly, but who now have better access to a better life including educational opportunities, health care and other essentials that include a voice in shaping their future.

Green innovation[1] often is perceived as the panacea to move our societies towards better times, greater prosperity, and with respect for nature. However, as we have learned on many occasions, innovation must be accompanied by deep and environmentally appropriate value systems. While such systems have gained ground, they are not robust enough yet. Especially when stacked against political decision making that favors short-term perspectives on problem solving, and entrenched special interests. What makes China such an important player for green innovation is that is attempting to build an ecologically based value system through its emphasis on ecological civilization; also that is has the means to act on a long-term vision. This may lead to significant shifts in values and capacity in other developing nations, whether through sharing of innovative technologies, improved access to green financing, and management skills to accelerate the pace of green development.

It will take a strong and combined sense of idealism and pragmatism to realize better global conditions for nature and humanity such as those well laid out in the UN SDGs and targets now being pursued at national levels. Acting strategically on these, and with full commitment sooner rather than later is the challenge of the century, since failure to do so has terrible implications for the longer term and certainly for the last half of the 21st Century.

China is well motivated towards idealism through its emphasis on ecological civilization and its commitment to a New Era. And there is a high level of pragmatism in the strengthening of environmental laws and regulatory enforcement, institutional reform including the new Ministry of Natural Resources and Ministry of Ecology and Environment, and many other actions. It is very encouraging that environment and the War on Pollution are now given high priority. It is also apparent that China's population continues to place emphasis on a clean and safe environment as a key element for a better life. We can be reasonably assured that, despite the magnitude of challenges, China through its domestic efforts is likely to progress well during its journey away from tipping points, towards turning points, and eventually to national environmental quality and ecological security.

Globally, China is positioned to build competitive and strategic advantages as it opens new trade and investment paths, and commits to development in many neglected or difficult parts of the world. It has made clear that it will not do so at

1 Various examples: innovation and green growth:http://www.oecd.org/innovation/inno/fosteringinnovationforgreengrowth.htm ; potential consumer-oriented innovations: (https://interestingengineering.com/21-sustainability-innovations-and-initiatives-that-might-just-change-the-world); Danielle Sinnett, Nick Smith , Sarah Burgess.green urbanization: Handbook on Green Infrastructure: Planning, Design, and Implementation. Edward Elgar Publishing. 2015; biodiversity conservation: http://www.trustforconservationinnovation.org/sponsored/

the expense of the environment. However, there are some skeptics on whether this is possible. This will clearly be one of the most important aspects of whether BRI and some other aspects of South-South cooperation are judged successful in the coming decade.

4.5 Realizing a Global Green New Era

As CCICED developed its Phase VI (2017—2021) research program considerable attention was given to the balance of activities focused mainly on domestic Chinese concerns and those addressing international issues. It is apparent that almost all activities now require a mix of both. Relationships are influenced by trade and investment policies including South-South cooperation and BRI, the rising ecological footprint of China, obligations under international agreements, and trans-boundary impacts such as those associated with global climate change and ocean sustainability. As well, Chinese environment and development research capacity has expanded quite dramatically over the past decade. At the same time, the issues have grown more complex, with the need for integrated policy solutions. These factors also influence action at the regional and global levels.

The short list of questions noted below is helpful for determining priorities in existing and future CCICED work, and advice concerning the challenges and opportunities for a global green new era.

(1)How can globalization be redefined along lines that respect Planetary Boundaries[1], and enhance Ecosystem Services[2] worldwide? The innovation required will demand new coalitions and continued capacity development within all nations. It will test our capacity to build bridges across political ideologies and create widespread participation opportunities for citizens—rich and poor, young and old, women and men. Will green job creation, sharing economy and other socially relevant actions help in bringing populist interests on board?

(2)How to buffer highly interactive environmental, socio-economic and political

1 Four of nine Planetary Boundaries have been transgressed: "The four are: climate change, loss of biosphere integrity, land-system change, altered biogeochemical cycles (phosphorus and nitrogen).Two of these, climate change and biosphere integrity, are what the scientists call "core boundaries". Significantly altering either of these "core boundaries" would "drive the Earth System into a new state." https://www.stockholmresilience.org/research/research-news/2015-01-15-planetary-boundaries---an-update.html

2 China is now one of the leading countries in the world trying to define the significance of its ecosystem services and protect them through ecological redlining and eco-compensation. See for example Yang Bai et al. 2018. Developing China' s Ecological Redline Policy using ecosystem services assessments for land use planning. https://www.nature.com/articles/s41467-018-05306-1

shocks while still continuing on a pathway of sustainable development at subnational, national, regional and global levels? Can new approaches such as ecological civilization materially help with this buffering, or open new pathways that avoid the shocks?

(3) How can synergies and integrated planning for land, water and ocean use, human settlement, biodiversity and ecosystem services accelerate progress on meeting critical 2030 SDGs and other essential targets and make action more cost-effective? This topic is well discussed, for example concerning linkages between global conventions on climate change, biodiversity protection and but action lags.[1]

(4)How can the digital economy contribute more effectively towards meeting the many challenges for developing sustainable production and consumption? Innovations such as applying block chain techniques related to the 2030 SDGs, green supply changes, advanced circular economy, and sustainable resource use, etc., are examples.[2] The concept of the 4th Industrial Revolution is another.[3] And the massive data banks of Google, Amazon, WeChat, Alibaba and others undoubtedly may lead to ways of changing consumer behavior towards sustainable development.

(5)How might China play a greater and sometimes leading role to bring about improved global environment global environment and development governance? Participants at the CCICED Brussels Roundtable[4] reviewed and generally agreed with a framework involving several key approaches:

Leading by example involves developing efficient domestic processes and policies and thereby setting ever more ambitious national goals. This will support establishment of new global norms supporting environmental global governance. China has done so when it comes to developing green finance and also to a certain degree when it comes to setting domestic climate targets (nationally determined contributions under the UNFCCC process).

Leading by providing resources can be done by providing financial support along the line China already is pursuing when it comes to climate change and South-South cooperation. Other ways of providing resources including capacity building,

1 CCICED held a Roundtable co-hosted with the EU Commission in June 2018. See transcript on Roundtable on Global Governance and Ecological Civilization; Roundtable Summary Report; and Discussion Paper Synergies for Improving Performance on Global Environment and Development Agreements.

2 https://blockchainhub.net/blog/blog/blockchain-sustainability-programming-a-sustainable-world/

3 https://trailhead.salesforce.com/en/modules/impacts-of-the-fourth-industrial-revolution/units/understand-the-impact-of-the-fourth-industrial-revolution-on-society-and-individuals

4 CCICED held a Roundtable co-hosted with the EU Commission in June 2018. See transcript on Roundtable on Global Governance and Ecological Civilization; Roundtable Summary Report; and Discussion Paper Synergies for Improving Performance on Global Environment and Development Agreements.

sharing of policy experiences and sharing of green technologies.

Leading by coalition building is about building clubs of similarly motivated countries seeking to drive the international political agenda and providing "good examples". Such clubs can be supported by linkages between different policy issues, e.g. seeking agreement on environmental issues through linkage to trade issues, cultural and educational exchange programs, etc. China's Belt and Road Initiative provides a huge opportunity for such linkages and hence building of coalitions that may support environmental global governance.

Leading by increasing the knowledge base necessary for enhanced sustainable development and ecological civilization progress. Some areas of particular significance include sharing of knowledge on impacts from development, especially from novel forms of ocean development, co-management in marine and terrestrial ecosystems, and on-going efforts to define and improve ecological services.

4.6 CCICED Research Strategy in Phase VI

CCICED's research activities are being undertaken via Special Policy Studies (SPSs) clustered under four Task Forces (TFs). Each TF has an emergent theme that will be explored over a period of years, informed by the more specific topics examined by the SPSs. There are eight SPSs at present, with three each in TF 1 and 2. TF 3 and 4 will have more SPSs added in the future. At present there is one in each. The selection of both TF themes and SPS topics and their work has touched on matters of high relevance to both China and the rest of the world. Titles of each are noted below along with a short description of emergent themes for each TF. The themes are subject to further elaboration and refinement during the coming year.

4.6.1 TF on global governance and ecological civilization

Emergent Theme: Acting on synergies among international governance agreements, focusing especially on the three SPS topics and on UN 2030 SDGs. Synergies should help to meet global, regional and national goals more quickly and comprehensively, achieve co-benefits, and provide greater resilience at an ecosystem and societal level.

(1)SPS on China's Contributions to Global Climate Governance
(2) SPS on Post 2020: Global Biodiversity Conservation

(3)SPS on Global Ocean Governance and Ecological Civilization

4.6.2 TF on green urbanization and environmental improvement

Emergent Theme: Integrated green regional development, while difficult, is essential to achieve comprehensive objectives and transformative changes for a Beautiful China and basic ecological civilization by 2035. This approach covers green urbanization with links to a revitalized rural, ecologically sound economy providing abundant ecological services. The Yangtze River Economic Belt is a very significant case, perhaps the most important in China.

(1) SPS on Green Urbanization Strategy and Pathways Towards Regional Integrated Development
(2) SPS on Ecological Compensation and Green Development Institutional Reform in the Yangtze River Economic Belt
(3) SPS on Goals and Pathways for Environmental Improvement in 2035

4.6.3 TF on innovation, sustainable production and consumption

Emergent Theme: Ecological civilization is based on green development principles being applied widely and consistently across all sectors. But these are not yet fully understood or consistently acted upon by either consumers or producers. With China's rise in incomes, the country's ecological footprint continues to rise, and consumers do not have sufficient information or opportunities to make green consumption choices. Green supply chains (domestic and international) will help to change behavior. Innovative green technology in China eventually may surpass what happens elsewhere in the world, given the substantial investment and huge potential markets.

SPS on Green Transition and Sustainable Social Governance

4.6.4 TF on green energy, investment and trade

Emergent Theme: China's financial investment at home and abroad must be consistently oriented to sustainable development. Green energy is a very important element. International trade agreements, whether bilateral of multilateral, should be made consistent with the 2030 SDGs, and provide for inclusion of environmental considerations. Much of China's investment and trade will be with G77 partners, and therefore special attention is needed on greening the Belt and Road Initiative, including sharing of Chinese environmental experience.

SPS on Green Belt and Road Initiative (BRI) and 2030 SDGs

In dealing with complex topics like those outlined above, CCICED expects to keep a strong focus on the underpinnings, specific topics and then examining the interactions needed to make rapid progress to green development and sustainable development/ecological civilization.

Underpinnings refer to the cross-cutting concerns that are common to many situations and initiatives. These include green finance needs, institutional issues, rule of law, and social matters such as gender mainstreaming.

The specific topics are matters such as appropriate green technology development, policy development for problems such as reducing the burden created by plastic debris entering the oceans, etc.

Interactions are the value added through systemic approaches to problem solving, for example, finding co-benefits from various types of circular economy in green industrial parks.

In strategic terms, this means extracting greater value at three levels in CCICED work. i) The insights provided from the individual SPSs. ii) The additional benefits of analysis within clusters. For example, the interaction between climate change and oceans And, iii) The ability to deal with green development in complex systems such as the YREB, or new urban-rural settings such as the state level Xiong'an New Area; and the Greater Bay Area (Guangzhou-Hong Kong-Shenzhen-Macau and other cities in and around the Pearl Delta).

These levels of analysis are intended to provide new insights into the issues China and others must consider in the pathways of sustainable development and ecological civilization. It will be challenging for TF co-chairs and SPS team leaders. We hope CCICED Members and Special Advisors will provide their expertise as well, for example in their inputs to the 2018 AGM Open Forums and in various roundtables.

4.7 Issues: Aiming High 2020—2035

In 2020 China will celebrate its achievements regarding a moderately well off Xiaokang Society free of the scourges of poverty and with much to showcase regarding environmental improvements. 2020—2021 is an important milestone on the way to constructing an ecological civilization by 2035. Getting there will be very difficult. China and many others must aim high. It is not a time to take small steps, or to hold back on action that can accelerate progress on green development. The time frame is only 15 years—only three Five-Year Plans for China's domestic green shift. And the time frame for important for major global transformations is even less, based on the SDGs, and the goals for climate change, etc. The issues noted below are among the most significant for a New Green Era that will be good for the planet and people everywhere. They relate to CCICED's work now and hopefully in the future. In the eight issues noted below views are slanted towards China's situation, needs and opportunities.

4.7.1 The 1.5℃ challenge and opportunity

The recent IPCC report on catastrophic climate change could not be more blunt. Boiled down to a single point it is that the world should "cut carbon pollution as much as possible, as fast as possible."[1] If we are unsuccessful in "bending the curve" of global warming over the coming decade, we face a dismal future. Aiming at a 1.5℃ limit is a must. China is not anywhere close, even with the rather herculean efforts over the past several years. Nor are other countries. It is time to lead by example.

The institutional reform in China's government can help to accelerate progress. For example, it will open up more co-benefits with the War on Pollution. Additional opportunities will emerge once green technologies are widely adopted. China is likely to be the leading source of electric automobiles and perhaps become the leading large country in terms of low carbon economy. This will require even greater incentives that exist currently. Enterprises of all types must be committed. Cities worldwide are demonstrating their interest in holding down global warming. Indeed if they do not, they are highly vulnerable to impacts. China has the great advantage that much of its urban infrastructure still remains to be designed and built. Green urban planning must play a larger role. China can transfer experience, technology and in some cases supply investment funding directed to stringent targets in other developing countries. There is need for stronger partnerships since no country on its own can do enough.

4.7.2 "Space for Nature"

The loss of biodiversity and ecological services continues throughout the world.

1 https://www.theguardian.com/environment/climate-consensus-97-per-cent/2018/oct/15/theres-one-key-takeaway-from-last-weeks-ipcc-report

The seeming inability to stem the losses despite well-intended plans nationally and globally presents a situation with the same degree of urgency as climate change mitigation. Nature has played second fiddle to short-term and long-term economic gain. Ecological restoration and biodiversity conservation is now needed on a grand scale. How grand it should be is an important question.

Some leading figures suggest we need to set aside half the planet to meet nature's needs, and half for human material needs.[1] This may appear to be a radical approach, but it may be equally radical to argue that all of the earth deserves to be managed for a strong respect for nature, with harmony between people and nature. The latter view might be taken for urban forests and constructed wetlands in cities. Or ocean space, which is three-dimensional, with much of its biodiversity found in deep ocean space, or carried in strong ocean currents between continents, such as whales and fish like bluefin tuna.

At present, China might well be close to half-protected status, with more than 15% of its land area in nature reserves, significant numbers of marine protected areas, and a large amount of its land and water area slated to be covered through functional zoning measures and ecological redlining. A lot depends on definitions, for example, does three month zero marine fishing areas in parts of China's seas qualify as protected space? Arguably, yes. The important point is that China is deeply engaged in protecting nature in the quest for an ecological civilization. The country will have the first stage of a national park system in place by 2020. It plans to review the status and condition of the 12,000 nature reserves.

By hosting the CBD COP 15 in 2020, China has an important opportunity to help set an agenda for action during the coming decade. This may be one of the most significant and timely opportunities for China to influence the outcome on the future state of the Planet's life.

4.7.3 Greening the Blue Economy

China's interests in the Blue Economy go from pole to pole and involve all maritime sectors. The capacity of China's ocean exploration involves icebreakers, submersibles in the deepest trenches, and very advanced satellite remote sensing. In China's EEZ and coastal zone there are serious problems of overuse, and conflicting priorities. Some areas such as the Bohai Sea and Yellow Sea need more attention to integrated, sustainable use management. China's distant water fleet fishes in many parts of the world, and the government currently is taking various

1 E O WILSON. Half-Earth. Our Planet's Fight for Life. 2017. http://books.wwnorton.com/books/Half-Earth/

actions to address IUU matters and other concerns such as fishing fleet subsidies. With modern technologies at hand, China can expect to play an important role in a new generation of maritime shipping, offshore mariculture, offshore energy production, and marine biotechnology.

Some Chinese estimates suggest that the Blue Economy could reach 30% of GDP by 2050 (currently it is about 10%).[1] For this to happen sustainably would require a green development strategy far beyond action taken so far. A case in point is the need to address plastics (including microplastics) entering the ocean from Chinese rivers. Ocean sustainability governance is complex, with numerous international bodies involved. It is one of the most significant areas for China to engage more in global governance and research.

4.7.4 Zero pollution urban strategies[2]

While it may seem unattainable, zero pollution must be an ultimate goal for at least some elements of the War on Pollution. For example, through the switchover to electric vehicles. And in the use of heat pumps or other natural sources of energy, including wind and solar power, and cooling water from lakes and reservoirs. Avoided material and energy uses and industrial waste utilization can be linked to resource efficiency and both circular and low carbon economies.

A recent UNEP Report[3] proposes stronger international action to tackle pollution. Five key messages are to:

i) exercise political leadership and partnerships to form a global compact on pollution;
ii) establish environmental governance policies for priority pollutants;
iii) develop a new integrated approaches for resource efficiency and lifestyle changes;
iv) mobilize finance and investment to drive innovation for new pollution control mechanisms;
v) create advocacy for action so that citizens and enterprises reduce their pollution footprint. A zero pollution approach in China can be aligned with support for new livelihoods, quality of life, and good health.

---

1 Tabitha Grace Mallory. Preparing for the Ocean Century: China's Changing Political Institutions for Ocean Governance and Maritime Development. Issues & Studies,2015,51(2): 111-138. https://www.researchgate.net/publication/303804064_Preparing_for_the_Ocean_Century_China%27s_Changing_Political_Institutions_for_Ocean_Governance_and_Maritime_Development.
2 http://www.resourcepanel.org/sites/default/files/documents/document/media/irp_china_case_study_policy_briefs_ramaswami.pdf.
3 Report of the Executive Director. 15 October 2017. Towards a Pollution-free planet. UNEP/EA.3/25http://web.unep.org/environmentassembly/report-executive-director.

4.7.5 Integrated river basin and coastal zone micro-level management

Fine-grained planning and management strategies can now be carried out utilizing new institutional mechanisms and a variety of technological tools. While river basin commissions and other bodies can be very helpful, in the case of China, they have not had enough clout or accountability (in contrast with the Rhine Commission or some other bodies). The new system in China of river and lake chiefs, and now bay chiefs brings environmental management accountability to very local levels. For agricultural chemical and water use there is considerable progress in reducing input levels via use of drones, and other means of remote sensing for environmental plans; and real time interventions to address the vexing issues of non-point pollution, unauthorized alterations of coastal lands, etc. Ultimately every mu of land should be assessed for ecological services, optimal sustainable use, and needs such as eco-compensation. Nothing short of a sustainability revolution will provide the social, economic and eco-environmental dividends needed.

The nature-oriented shift in use and ecological restoration underway in the YREB will benefit from a mountain to sea approach that covers all types of ecosystems and consideration of how they are linked and related to water uses. Hopefully the experience from integrated management in this very large and very complex system can be transferred to other parts of China, and for management of boundary waters.

4.7.6 From environmental protection to ecological civilization 2035

In recent years globally and in China we have seen that many older institutions and ways of doing things may hold less relevance to a new generation. Knowledge takes on less relevance as it becomes part of obsolescence, and sometimes even as it contributes to major improvements. Marketing shifts to disposable goods, with replacement even before the expected period of usability is reached. These and other trends can play out at all levels from individuals and households, to national and global situations. Turmoil can prevail, especially when governance lags behind the rate of impacts. There are numerous examples. One is the difficulty of implementing carbon taxes, even though conceptually rationale. Another is the perception of environmental risks—assessments of risk level often are much lower if it involves a personal decision compared to risk associated with governmental decisions. And a third is the well-known NIMBY (not in my backyard) phenomenon.

What can be promised today for times 15 years hence? Will people be addicted to automobiles as much as they are today? Will environmental health impacts rise even as pollution levels drop, given the aging populations in China and other countries? Will China's material well being be satisfied at much lower levels than in many western societies? Will emerging technologies provide net benefits for green development or mainly introduce new challenges for the existing problems? These and many other questions make predictions about 2035 difficult—and the situation for another longer-term milestone (2050) is even more uncertain. China has demonstrated that its target setting has produced many benefits over the past decades. It helps to build a degree of resilience as well.

A roadmap of key policies for the transition from a sector-based approach of environmental protection to a more comprehensive ecological civilization approach provides for additional perspectives. Generally, however, some of the economic models and other predictive tools that have been used in the past are too linear for the uncertain times ahead.

4.7.7 Green supply chains

The foundation of modern globalization, supply chains can be a means of ensuring good practices, of building capacity, of rapidly spreading the fruits of technology innovation, and for providing assurance to both producers and consumers that products are safe and produced with proper consideration of the environment. This is an idealized vision of supply chains—where standards are robust and transparent in their application, and where a strong motivation exists all the way from raw materials to final consumer use and disposal. Environmental considerations should enter into each stage. Reality is often very different. Commodity supply chains are often in the spotlight, especially for situations that involve land conversion (palm oil, soy beans), uncertain chain of custody (distant water fishing fleets), illegal sourcing, failure to take into account circular economy considerations (disposal of electronic wastes), etc. Increasingly there is concern to ensure that carbon footprints are included, and undoubtedly other considerations such as impacts on the oceans of plastic products is addressed via supply chain controls.

Greening of supply chains is often seen to be in the domain of multinational firms that depend highly on maintaining their social license to produce and/or sell to consumers. Companies such as Walmart and IKEA and many others have made a genuine effort to make green supply chains an important part of their business strategy. And being part of a long supply chain can allow smaller enterprises to make or market products internationally. These small and medium-size enterprises may struggle to meet green standards without considerable assistance. And there can be differences between national and international standards. Such problems hinder the global progress towards green supply chains. Efforts such as "Made in China 2025" could be helpful in encouraging widespread adherence to sustainability standards. Consumers need the help of enlightened marketers to provide reliable sustainability evidence and a broader range of green products. The huge consumer markets of China can influence the rapid introduction of new green product types while lowering the cost for everyone in the world. Illumination using LEDs is an example.

4.7.8 Linking Belt and Road with 2030 SDGs

The current reach of the BRI is focused mainly on infrastructure improvement in partner countries, sometimes on a massive scale (e.g., in Pakistan) However over time it undoubtedly will include other components of trade and investment, some of which might accelerate the pace of meeting their 2030 SDG action plan targets. The value of this linkage is that the results will be linked to national priorities; and for some matters such as climate change, or ocean sustainability will also contribute to global goals. There may be significant green opportunities, and also the potential to build a better understanding of ecological civilization. It would be helpful to build a 2035 green BRI strategy that takes into full account 2030 SDGs, and to periodically compare this with actual achievements over time in various countries, including China's own progress.

This approach would be in line with China's domestic efforts for 2035 environment and ecological progress. It also could be helpful in the reform of China's own development cooperation policies, South-South cooperation on climate change and various regional concerns such as green development in the Greater Mekong Subregion.

4.8 Conclusion

This Issues Paper started with consideration of shocks and new eras. Many more could have been introduced, for example, the impacts of global demography with aging population in countries such as China; and countries populated by predominantly young people. Certainly it will be hard to reconcile the intergenerational differences in needs, and perhaps attitudes towards green

development. Some experts predict a substantial shock in the form of climate change refugees, and with that possibility, civil unrest and conflicts over land and water. Health and the environment will remain a major concern, especially around national and local capacity to deal with such matters. China has wisely taken this on board as a major political concern, although much more remains to be done—a major theme certainly to 2035. Other countries in Asia, including Indonesia, Vietnam and cities in India now face their own crises in air and water pollution. To what extent can China help by sharing its experience?

All these examples point to the essential role social and political dimensions will play in determining sustainable development and green civilization outcomes. The emerging New Green Era must pay great attention to these aspects, and build a positive dialogue around new job creation, healthy living in a clean environment, and efforts to narrow income gaps, including the social preparations for making a good living either in town or countryside. Of course transparency and opportunity for all people to contribute to green conditionality in their personal decisions and societal progress is essential. This makes gender mainstreaming so important in all countries. By any measure, we will not see full advancement to an ecological civilization until there is expanded recognition of the role non-state actors can play, whether as community leaders, business leaders, or as committed institutions that can provide R&D support, and those athletes, entertainers and many others, who change the public attitudes due to the respect and attention given to them by ordinary citizens.

Undoubtedly China will be a torchbearer for a green and sustainable world in this time rife with contradictions that threaten our future. Hopefully many others will continue to support such efforts.